Inclusive Pathways for Marine Resource Stewardship and Global Prosperity

Ali Gökhan Gölçek
https://orcid.org/0000-0002-7948-7688
Niğde Ömer Halisdemir University, Turkey

Bilal Göde
https://orcid.org/0000-0001-8377-5909
Pamukkale University, Turkey

IGI Global Scientific Publishing
Publishing Tomorrow's Research Today

Vice President of Editorial Melissa Wagner
Director of Acquisitions Mikaela Felty
Director of Book Development Jocelynn Hessler
Production Manager Mike Brehm
Cover Design Jose Rosado

Published in the United States of America by
IGI Global Scientific Publishing
701 East Chocolate Avenue
Hershey, PA, 17033, USA
Tel: 717-533-8845 | Fax: 717-533-7115
Website: https://www.igi-global.com E-mail: cust@igi-global.com

Library of Congress Cataloging-in-Publication Data

Names: Golcek, Ali Gokhan, 1990- editor | Göde, Bilal, 1990- editor
Title: Inclusive pathways for marine resource stewardship and global prosperity / Ali Gökhan Gölçek, Bilal Göde, editors.
Description: Hershey, PA : IGI Global Scientific Publishing, 2026. | Includes bibliographical references and index. | Summary: "This book aims to address the sustainable management and utilization of marine resources from a global welfare perspective"-- Provided by publisher.
Identifiers: LCCN 2025021049 (print) | LCCN 2025021050 (ebook) | ISBN 9798337317779 hardcover | ISBN 9798337317786 paperback | ISBN 9798337317793 ebook
Subjects: LCSH: Marine resources conservation | Marine resources development
Classification: LCC GC1018 .I55 2026 (print) | LCC GC1018 (ebook)
LC record available at https://lccn.loc.gov/2025021049
LC ebook record available at https://lccn.loc.gov/2025021050
British Cataloguing in Publication Data
A Cataloguing in Publication record for this book is available from the British Library.

Table of Contents

Detailed Table of Contents

An inclusive approach to creating sustainable entrepreneurship in the marine sector is essential, with human resources as a strategic factor. Amidst increasing environmental challenges, sustainable management of marine resources requires green entrepreneurship. Human resources play a role in driving innovation, strengthening local capacity, and ensuring the involvement of women, youth, and indigenous communities. Policies that support skills education, economic empowerment, and access to green technology are needed. Collaboration between government, the private sector, academia, and local communities will create sustainable business models that support SDG 8 (Decent Work and Economic Growth) and SDG 14 (Marine Ecosystems).

Marine tourism, one of the fastest-growing segments of global tourism, benefits from natural, cultural, and economic resources, especially in coastal areas. It includes diving, water sports, yacht and cruise travels, and whale watching (Hall, 2001), attracting diverse tourist profiles and supporting destination promotion. However, besides economic gains, it also brings sociocultural and environmental impacts. Mass tourism may lead to ecosystem degradation, sea pollution, and loss of biodiversity (UNEP, 2009). It can also affect the local community's economy, culture, and traditions in both positive and negative ways. This study aims to contribute to the literature by examining these ecological and social effects. The first part discusses the concept, development, and importance of marine tourism, while the following sections evaluate its environmental and social impacts.

Climate change (CC) and the increasing occurrence of extreme weather events present significant threats to coastal regions, particularly those reliant on tourism. This chapter analyses the case of the Mediterranean Costa del Sol in southern Spain, a destination where tourism revenues and attractiveness are closely tied to the environmental quality of its coastline. The study investigates beach users' willingness to pay (WTP) to support conservation and adaptation measures aimed at countering CC impacts. Findings show that visitors are generally willing to contribute financially to beach conservation, though this disposition is shaped by socio-economic and attitudinal factors, including income, level of education, and environmental concern. For coastal municipalities, the central challenge is to transform this willingness into practical financing mechanisms that are socially accepted, transparent, and equitable. This study provides critical evidence for designing inclusive coastal policies and awareness strategies to strengthen the climate resilience of beach destinations.

This study analyzes future water stress trends in Turkey and identifies the key socioeconomic and environmental drivers shaping this trajectory. Using data from 1992–2022, a multivariate polynomial regression model was developed with sectoral water use indicators, population, economic growth, CO_2 emissions, and drought measures. The model effectively captures nonlinear dynamics and explains most of the variation in water stress. Results show that population growth, agricultural withdrawals, rising GDP, and CO_2 emissions intensify water stress, while water use efficiency indicators exert a mitigating influence. Projections suggest that Turkey's water stress will continue to rise through 2050, potentially reaching critical levels under current trends. The study underscores three policy priorities: improving agricultural efficiency, reducing urban water losses and expanding reuse, and transforming water-intensive industrial processes. A holistic water management framework is essential for strengthening Turkey's resilience to escalating water stress.

Preface

Marine and ocean ecosystems are critical to our planet's climate stability, biodiversity, and societal well-being. However, factors such as climate change, pollution, overexploitation, coastal pressure, and inequality in benefit sharing are making the sustainability of marine resources increasingly fragile. Inclusive Pathways for Marine Resource Stewardship and Global Prosperity discusses how to design inclusive pathways that bring together science, policy, finance, technology, and community expertise in these fragile areas.

Inclusivity is more than just seeking stakeholder input. It requires placing local communities, women, youth, small-scale fisheries, SMEs, and indigenous/local knowledge at the center of decision-making processes within a rights-based framework. This includes considering the equitable sharing of benefits and risks, traceability and accountability, investment in human resources, and combined tools such as skills transformation and green employment. This book demonstrates, based on concrete examples and evidence, the conditions under which practices that embrace the growth orientation of the blue economy, as well as the goals of ecological integrity and social justice, can be successful.

To this end, the chapters: These include artificial intelligence and digital oceans, blockchain, fintech tools, critical infrastructure security, human capital-based institutional transformation, marine tourism, ecotourism, willingness-to-pay (WTP) studies for coastal protection, and climate-water economics in the Turkish context. The shared goal is to develop scalable, applicable, and community-based solutions: traceability and combating IUU, data-driven environmental management, emission reduction in maritime shipping, the balance between ecological carrying capacity and community benefits in tourism, inclusive business models, and the economic implications of climate risks.

ORGANIZATION OF THE BOOK

Integrating Ethics, Innovation, and Inclusivity in Maritime Governance: Toward a Regenerative Blue Civilization: This opening chapter lays out the book's foundational framework by outlining a vision of a regenerative blue civilization. It holistically addresses ethical values, inclusive governance, innovation, and long-term ocean management perspectives. The chapter provides a strong theoretical foundation by discussing the future of regulatory frameworks, the integration of social justice into the blue economy, and the principles of a new maritime civilization.

AI-Enabled Marine Intelligence and Community-Centered Stewardship: Digital Oceans & Equitable Blue Governance: This chapter comprehensively examines how digital oceans and artificial intelligence are transforming marine governance. It details how data analytics, sensor networks, remote sensing, and community-based governance tools are converging to strengthen both ecological monitoring and decision-support systems. The authors demonstrate how technology can be integrated with community participation, justice, and accountability.

Cybersecurity in the Water Economy: Threats and Policies for Critical Infrastructure: In a water and maritime economy increasingly digitized, the cybersecurity of critical infrastructure is one of the fastest-growing global risks. This section analyzes the threats facing water infrastructure, maritime operations, and data-based management systems, while offering actionable protection strategies for policymakers. It also evaluates technical and administrative frameworks for enhancing cyber resilience.

Blockchain in Conserving Marine Resources and Combating Illegal Fishing: This chapter demonstrates blockchain's potential in combating illegal, unreported, and unreported (IUU) fishing. Tools such as supply chain traceability, product de-duplication, smart contracts, and IoT-enabled tracking systems are systematically discussed. The challenges of data integrity, privacy, and interagency collaboration are supported by case studies.

Integrating Green Technologies in Maritime Transportation With Deep Learning: Centered on the use of deep learning and green technologies in maritime transportation, this chapter explores next-generation analytics for route optimization, fuel efficiency, predictive maintenance, and emissions reduction. A roadmap for sectoral transformation is presented, encompassing digital twin applications, port integration, and compliance with international regulations.

Inclusive Pathways to Sustainable Marine Enterprises: The Role of Human Resources: This section discusses the human capital structure that underpins inclusive maritime businesses. Topics such as skills redefinition, upskilling/reskilling, inclusive employment, community-based entrepreneurship, and corporate culture transformation are explored through real-world examples. It explains how education, public, and private sector collaborations can be institutionalized.

Marine Tourism: Niche Segments, Ecological Challenges, and Community Impacts: This section focuses on the ecological risks and socioeconomic impacts on communities of niche marine tourism areas such as diving, yacht tourism, and cruise routes. Recommended policy tools for balancing carrying capacity, waste management, environmental sensitivity, and community benefits are presented comprehensively.

Climate Change and Willingness to Pay for Beach Conservation: Insights from Costa del Sol (Spain): This chapter analyzes societal preferences for financing coastal protection through a WTP study conducted in the Spanish context. The impacts of climate risks on coastal erosion, infrastructure pressure, and tourism revenues are assessed using quantitative methods. Equitable financing models for different income groups are discussed.

Türkiye's Water Future: Demand Forecasting and Sustainable Strategies / The Relationship Between Climate Change and Marine Economics in Türkiye: These chapters, based on the Turkish context, analyze the impacts of climate change on water resources, the maritime economy, and sectors. They delve into demand forecasting models, sustainability strategies, the economic implications of climate risks, and policy options. This national assessment strengthens the book's global framework with regional data.

We extend our gratitude to all our authors, reviewers, and institutions who contributed to the creation of this book. We hope that the suggestions here will serve as concrete guidance for policymakers, practitioners, and researchers who view the oceans not only as an economic space but also as a life-support system and shared heritage. We also extend our sincere gratitude to the IGI Global team for their professional support at every stage of the publication process.

Chapter 1
Integrating Ethics, Innovation, and Inclusivity in Maritime Governance:
Toward a Regenerative Blue Civilization

R. Vettriselvan
https://orcid.org/0009-0005-1097-0806
Academy of Maritime Education and Training, India

ABSTRACT

This chapter explores how gender equity, circular port management, and climate-resilient innovation collectively redefine the foundations of the blue economy. It examines how inclusive participation, sustainable infrastructure, and intelligent technologies such as AI and blockchain foster transparency, adaptability, and regeneration in maritime systems. By linking social justice with environmental and economic performance, the study highlights the transformative potential of gender-inclusive governance, circular logistics, and predictive climate strategies. The chapter envisions a regenerative blue economy where equity, technology, and ecological stewardship converge to promote sustainable prosperity and resilient global maritime development.

DOI: 10.4018/979-8-3373-1777-9.ch001

1. INTRODUCTION

The dawn of the twenty-first century has seen the radical shift in the maritime sphere where oceans, previously perceived as sources of resources, as well as as routes of trade are currently considered to be the key to sustainability, advancements in technology, and the overall economic revolution. This shift is reflected in the changing paradigm of the blue-economy that views marine and coastal resources as the sources of interrelated goals of economic prosperity, social equality, and environmental health (Alhowaish, 2025; CCICED, 2025). Unlike the previous development models that focused on extraction and throughput, the modern approaches project the concept of integrating gender equity, circular port management, and climate-resilient innovation in one scheme and positioning profitability and long-term ecological stewardship. With countries trying to balance the exploitation of the seas and their conservation, the nexus of social inclusion, circularity, and technological resilience can be a key feature of sustainable maritime governance. The opportunities and constraints are increased by the global growth of blue-economy sectors. Shipping, aquaculture, coastal tourism, offshore energy, and marine biotechnology are some of the industries that generate trillions of dollars to the global economy and provide over 350 million livelihoods (Mehrotra, 2025). However, the same growth curve compelled the worsening of the ecosystems, overfishing, and habitat loss that now pose a threat to the persistence of economic growth and environmental balance (Batool et al., 2025). These paradoxes highlight the imminence of a change in the paradigms of traditional forms of growth into more regenerative, inclusive, and innovation-driven modes. In this regard, new governance systems require ethical vision, technological wisdom, and participation in ocean governance (Al Mokdad, 2025). This transition has also become important on the aspect of social inclusion, especially gender equity. Although women, youth, and marginalised communities have very important roles to play in fisheries, logistics, and coastal conservation, the representation of these groups in the structures of decision making is significantly low (Cavole et al., 2025; Williams, 2023). Gender equity is being seen as a key strategic facilitator of sustainability, which creates institutional trust, innovation potential, and resilience in maritime systems. It is established in research that inclusive governance leads to policies that are more adaptive and more responsive to the community, particularly in fisheries and coastal management contexts (Nnanguma, 2024; Vettriselvan et al., 2019). Therefore, gender inclusion does not only have a moral imperative but it is a structural pre-condition of blue-economy governance.

In line with these social imperatives, the logic of operation of maritime logistics worldwide is being transformed by circular port management. The ports that were traditionally structured around the linear flows of resources are moving towards the closed-loop models with the focus on reuse, renewable energy, optimised logistics, and digital monitoring (Cunha et al., 2025). Circularity allows ports to minimise externalities to the environment and promotes the symbiosis of industries with waste valorisation and energy efficiency. These transitions are further reinforced with the help of digital tools like AI-powered analytics and blockchain-based transparency to facilitate predictive maintenance, the tracking of emissions, and responsible reporting (Raheja et al., 2025; Babu et al., 2025). All these practises are examples of how the principles of the circle can help enhance the speed of decarbonisation and make global supply chains more resilient. The third main dimension of the modern ocean governance is climate-resilient maritime innovation. Risks to coastal infrastructure, fisheries, and livelihoods Due to sea-level rise, acidification, and storms of increasing intensity and intensity, there has been a significant threat to coastal infrastructure, fisheries, and community livelihoods (Goodman et al., 2023). Maritime industries are resorting to adaptive technologies and governance to ensure the threats are mitigated. The environmental forecasting, marine spatial planning, and emergency response have become the focus of AI, IoT-based sensing networks, and blockchain applications (Al Subhi & Al Suqri, 2025; Sultana et al., 2024). Such innovations underline the appearance of marine techno-ecology when technological instruments and ecology systems develop together to make decisions based on data that ensure the safety of the environment and the continuity of operations (Gesami 2024).

Equity and circularity and innovation are interwoven to create a networked system that has been defined by recent scholarship as a regenerative blue-economy. This change redefines sustainability as an external factor to a maritime competitiveness indicator and institutional legitimacy. India, Kenya, Greece, the Gulf States are developing national blue-economy plans which are more often focused on social inclusion, renewable ocean power, and AI-based environmental controls (Alhowaish, 2025; Ekpemuaka et al., 2025; Papadaki et al., 2025). International organisations like the UN Sustainable Development Goals (especially SDG 14) and the decarbonisation agenda of the International Maritime Organisation also contribute to the necessity of more ethics, economics, and technology-based governance models (Mutambara, 2025; Zaeri, 2025). The chapter also fits into the developing discourse of providing an analytical synthesis of the ways the principles are united to promote resilient, equitable, and innovation-driven maritime development. The plurality of ocean industries must be transformed to inclusive and climate-aligned systems by reworking of institutional and market setups. Following the recommendation of Al Mokdad (2025), sustainable governance should be seen not only as a compliance with regulations but also as intelligence-oriented decision frameworks and participative structures. Additional complementary results by Bakker (2022) also indicate that AI and real-time analytics can be used to improve conservation results and outcomes in marine protected regions in addition to complementing local livelihoods. These lessons show that maritime innovation is not simply technological but relational, it has an impact on the governance practises, forms of cooperation, and social justice. Inclusive leadership together with circular infrastructure and adaptive technological tools can provide opportunities to reshape the maritime industry as a recreation ecosystem. The change will guarantee environmental resilience, economic diversification, and enhanced social stability as an investment in resilient coastal systems will create jobs and facilitate work across borders (Le Gouvello & Simard, 2024). Re-forming a regenerative blue-economy is therefore a moral obligation as well as a strategic necessity to ensure the world is a prosperous place to live.

2. GENDER EQUITY IN THE BLUE ECONOMY

In coastal areas, women form a considerable but grossly underestimated portion of the workforce in the sea. They are involved in fishing, aquaculture, coastal tourism, seafood processing, and community management, often on a low-paid and informally and unpaid basis. According to Cavole et al. (2025), women do almost half of all small-scale fisheries operations in the world, but only a small portion of large-scale leadership or management. This difference indicates deeply rooted gender inequalities in the economic and cultural systems, which still hide the economic worth of women labour and limit their resources, financial and technological resources. Since the blue economy aims at fostering sustainable progress by using marine resources, the continuation of such inequalities goes against the social justice, as well as the sustainability of the sector in the long term. Gender equity thus serves as more than a moral issue but rather an economic issue and governance issue as well. It has been proven that under the conditions of meaningful involvement in fisheries management, marine solidarity, and maritime entrepreneurship, communities are characterized by the enhanced sustainability levels, uplifted household well-being, and adequate management of the resources (Cavole et al., 2025; Williams, 2023). Inclusion then comes to be an aspect of adaptive and resilient governance. International bodies like the IUCN and UN Women argue that involvement of women in value chains of the blue-economy will boost innovation, mitigation of risks and environmental standards. This claim is supported by the empirical evidence provided by Pacific and African fisheries: societies that have a higher level of female representation in the governance system are more likely to practise conservation and have more livelihood resilience (Handmer et al., 2024; Nnanguma, 2024). The inclusion of gender where it is institutionalised by decision making which participates, leads directly to social legitimacy and stability of the ecosystem further supporting the overall objectives of sustainable marine management. Structural blocks are widespread even with this international acclaim. Women are limited in their movement in the maritime industries due to patriarchal norms, unequal inheritance and property rights, as well as limited access to education. As Williams (2023) points out, financing systems rarely use gender-responsible criteria, thus women entrepreneurs cannot access capital when operating in new business areas (e.g., eco-tourism, aquaponics, or maritime logistics). The ability to overcome these restraint is compensated by occupational segregation that places women into low-income processing and marketing jobs and denies them any technical, managerial, and scientific jobs. Similar patterns are observed in the Indian technical industries by Vettriselvan et al. (2019), as gender diversity is positively related to productivity and innovation. As a result, women are being comparatively underrepresented in technology-intensive areas of the maritime sector, which hinders the capacity of the maritime industry to deal with issues creatively and reduces the institutional capacity during crises.

These inequalities are enhanced by institutional weaknesses in the marine governance. Most regulatory frameworks do not have the gender-disaggregated data which makes it harder to identify the magnitude of women input and evaluate the policy impact correctly (Medhekar, 2025). In the few cases where gender policies have been put in place, such policies usually tend to be rhetoric unless they are backed up with specific budgets, indicators, and cross-ministerial coordination. Consequently, policies become disjointed and incomplete, as gaps are indicated between policy aspiration and the capabilities of operation. It is essential to empower gender equity by therefore establishing gender considerations within the governance systems at all levels in policy process, such as planning, budgeting, monitoring, and evaluation. Al Mokdad (2025) is keen to point out that inclusive and smart governance should take into consideration integrating digital tools with participatory processes to become more transparent and representative. In the maritime setting, it implies the incorporation of gender analytics into the processes of marine spatial planning, port development, and resource distribution. An initiative that promotes gender-inclusive stakeholder mapping is supported by the Marine/Maritime Spatial Planning Guide (Iglesias-Campo et al., 2021), which can be implemented with a specific leadership training programme, gender-oriented budget, and the creation of participative councils in maritime organisations. Accountability can also be enhanced by utilising digital platforms and AI-driven decision support systems (Santos and Carvalho, 2025), which can also help monitor the inclusion metrics and make sure that gender commitments are converted into tangible results.

These are practical examples of such ideas in the form of regional initiatives. Women-owned cooperatives operating in the Kenyan coastal fisheries have implemented AI-based value-chain optimisation devices to increase the market transparency and fair pricing (Gesami, 2024). On the same note, in the Philippines and Indonesia, the female-led conservation networks use the decentralised digital governance systems to govern community reefs, hence strengthening biodiversity and participatory governance effects. These instances indicate that when combined with the digital innovation, gender mainstreaming helps to enhance the effectiveness and credibility of the marine management systems. Financial inclusion is another condition of equity in the development of a blue-economy. The conventional funding streams such as government grants, foreign loans and domestic investments favoured large industrial players thus marginalising micro- enterprises and women cooperatives. Further expansion of access by using blockchain-based credit rating and micro-investment pooling can be achieved through FinTech mechanisms discussed by Kaur, Kapur, and Kurucz (2025). The development of new models of impact-investment in Global South are paying greater attention to gender parity in funding requirements; as an example, blue-economy financing programmes in the Western Indian Ocean insist that a significant share of their fundee portfolio be female-owned. Microfinance systems based on blockchain also enable checking the environmental performance through smart contracts and, consequently, help women-owned coastal enterprises to attract sustainable investment (Khan et al., 2025; Raheja et al., 2025). Gender-responsive financing is not only effective in capital but also in reinforcing social-environmental value of blue businesses, which is associated with SDG 5 and SDG 14. Financial innovation and regional collaboration can together promote the objectives of diversification and gender equality in the Gulf region, as exemplified by Alhowaish (2025), when prioritised between the national blue-economy plans.

As a means of continuing progress, education, leadership development, and capacity building are crucial. Cavole et al. (2025) note that the underrepresentation of women in the field of ocean science and maritime engineering is supported by the lack of mentorship frameworks and stereotypes. The idea of creating extra specialised courses in maritime institutions, combined with leadership mentoring programmes, can develop a new generation of women in leadership who will be capable of handling technical and governance issues. Women are already playing a crucial mediating role between scientific institutions and traditional resource users already in many communities along the coasts of the country, a phenomenon that reinforces the hybrid governmental forms that are already advocated in order to develop sustainable fisheries management (Outa et al., 2021; Handmer et al., 2024). The gender inclusion also increases climate resilience in the maritime systems. Rarely are women the owners of resources like water, food and coastal vegetation which are quite susceptible to climatic changes. Their personal experience is the basis of the early warning systems to make adaptation planning more inclusive (de Heer & Choudhury, 2025). The process of designing AI and remote-sensing tools, in collaboration with women groups, to monitor shoreline erosion, storms, or habitat restoration is important to make sure the technological solutions are context-sensitive. Participatory projects in Bangladesh and the Pacific Islands, which use AI-based climate models, have both enhanced disaster preparedness and empowered women as data interpreters and community educators (Harrington,

2022). These illustrations are emphasized on the argument that gender inclusion does not necessarily only complement marine innovation but also enhances its social validity and functioning. Finally, to promote a gender-equitable blue economy, it is necessary not to look at women as passive recipients of blue economy benefits but as co-builders of maritime change. A shift in the paradigm like this requires a framework of governance structures that are able to entrench the element of inclusion by establishing enforceable laws, specific funding and institutional accountability. Regenerative blue economies are about strengthening the connexions between the social and ecological systems, as Le Gouvello and Simard (2024) claim. The role of women in this regenerative logic is key in that innovation is kept within the priorities of the community and that the maritime development is carried out in a manner that is resilient, equitable and future-ready.

3. CIRCULAR PORT MANAGEMENT AND DIGITALIZATION

The development of the modern ports is a paradigm shift: these ports have ceased being a mere logistic point to becoming an active driver of sustainable innovation. In the first half of the twentieth century maritime logistics was based on a linear 'take-make-Dispose' pattern where throughput efficiency is favoured over ecological regeneration. This kind of paradigm has contributed materially to the carbon emission, accumulation of waste material as well as the degradation of the coastal ecosystem. The rise of global decarbonisation agendas and Sustainable Development Goals has placed ports on the forefront of the movement towards circular port management (CPM), a structure that predicts the recirculation of resources and waste reduction, renewable energy and digital intelligence (Cunha et al., 2025). The essence of CPM, however, aims at operationalising closed loop logistics in which there are reuse, re-processing or reuse of materials and energy and not discarding. This orientation is consistent with the regenerative logic of the blue economy and views the ports to be central to the development of climate neutrality and socio-economic well-being of communities living around them (Alhowaish, 2025). Since ports deal with 8090 percent of the world merchandise in terms of weight, their management carries significant consequences of global sustainability paths. The case of investment flows in the Greek ports helps to show how the approach of integrating circular investment principles in the socio-economic decision-making process can produce the corresponding returns in terms of the environment and finances (Papadaki, Halkos, and Koundouri, 2025). Circular port management can tune shipping emissions, coastal pollution, and urban development, and can achieve it by actions, such as increased energy efficiency in terminal operations, electrification of cargo-handling equipment, use of alternative fuels, such as green hydrogen or ammonia, and installation of waste-to-energy systems, which serve local residents. Most of these action plans are aligned to Green Port Certification programmes and EcoPort frameworks, and environmental performance is aligned to ISO 14001 standards. The general objective is to design strong port infrastructures, which are energy-neutral, with microgrids of renewable energy sources, and technologically connected with smart logistics systems. This kind of development is an illustration of increased interdependentness between land-based urban sustainability plans of smart cities and new models of maritime circularity (Martinez-Mireles et al., 2025).

Digital transformation is one of the key levers that can drive this transformation. Intelligent circular governance has its cybernetic basis provided by artificial intelligence (AI) and blockchain. By using AI-based predictive analytics, it is possible to reduce fuel consumption, optimise the process of berth assignment and predict equipment malfunctions, which would make operations more efficient. In turn, blockchain supplements traceability and transparency through the value chains by creating unalterable records of the environmental compliance and a circularity measure (Raheja et al., 2025; Babu et al., 2025). Smart contracts can help in sales documentation of waste recycling, verification of carbon credits, and authentication of sustainable procurement. The interaction between AI and blockchain, as Suresh Babu, Begum, and Karafolas (2025) argue, is resulting in a new paradigm, in which the decentralized accountability models and machine-learning-driven sustainability models reinforce each other. Singapore, Rotterdam and Los Angeles, among the largest ports, have adopted the idea of a digital twin, which is a digital representation of the port infrastructure, to model the logistics flows, detect the environmental bottlenecks, and optimise the decision-making processes in operations. To reduce the resources inefficiency, these systems combine IoT-based devices that capture environment-related data, emission-monitoring devices, and mechanised waste-sorting systems. Similarly, AI-based decision dashboards help port administrators to make a trade-off between trade performance and environmental requirements (Santos and Carvalho, 2025). Industrial symbiosis is another important aspect of the circular port design whereby waste or by-products in one process are used as inputs in another. Such examples are the reuse of ship-dismantling residues, treatment sludge of a ballast-water processing plant, or bioenergy production based on organic waste production in port facilities. Antwerp and Hamburg are the first ports to establish eco-industrial areas, where terminals are located and whose logistics networks are optimized by AI to allow resources to exchange (Elston et al., 2024). The ability of circularity to create new economic value is evidenced in waste valorisation, in which the discarded materials become marketable products. Bio-based port economies, where the algae or marine biomass can be converted into biofuels or bioplastics can also be seen as the sources of prospective innovations (Sultana et al., 2024). The said bio-industrial pathways do not only restore the local ecosystems but also widen the job opportunities especially to the youth in the coastal areas and women hence solidifying the inclusion agenda as earlier discussed.

A shift in the circular port management requires a significant amount of money. Traditional models of financing, which focus on short-term throughput are usually at odds with the long payback periods of circular infrastructure. According to Kaur et al. (2025), the mobilisation of investors and the incentive of certified circular initiatives can be achieved through FinTech-based sustainable finance systems through transparent project verification via blockchain. Circular port development is more and more being underwritten through public-private partnerships (PPPs) and green bonds, blue bonds, and sustainability-linked loans. Al Mokdad (2025) notes that such financing might be made more effective with the help of intelligent governance structures, which should coordinate the flow of investments with policy incentives and provide the equitable allocation of benefits. In the developing economy, microfinance systems based on ledgers and decentralised allow the smaller coastal businesses such as fishing cooperatives and local recyclers to engage in circular port activities (Khan et al., 2025). This inclusion widens the economic involvement and prevents marginalisation of ports which are technologically supported. The greater supply chain is also restructured by circular port management. The conventional maritime logistics are often characterised by disjointed information flows between shippers, the customs agencies and the terminal operators. Through combining AI predictive logistics with blockchain-protected electronic records, ports will be able to achieve better synchronisation, minimise delays, and emissions. Gesami and Nunoo (2024) show that AI-enhanced cold-chain logistics of Kenyan fisheries by coastline can reduce spoilage and energy waste, which can be generalised to the maritime cargo processes. Predictive analytics may also be used to match cargo arrival and renewable energy, thus reducing the tendency to rely on fossil fuels. Reverse logistics as a vital circularity mechanism can also be streamlined with the help of machine learning that will map out the most optimal possibility of reuse or recycling returned materials (Mutambara, 2025).

Effective policy integration between maritime, municipal and regional authorities is a requirement of successful transitions of circular ports. Alhowaish (2025) highlights that circular innovation cannot work in Arabian Gulf because of ministerial fragmentation, which needs to be replaced by coordinated governance. The regional initiatives are represented by regional bodies, like the International Maritime Organization (IMO) and the Blue Economy Council of ASEAN, which facilitate the harmonisation of the regulations and standardised sustainability indicators. Circularity is incorporated in the policy instruments by the Extended Producer Responsibility (EPR) to ship waste and Zero-Emission Port requirements, which are included in legal frameworks. When it comes to addressing the environmental protection in conjunction with digital transformation, Ekpemuaka et al. (2025) state that an African blue economy relies heavily on bridging the digital transformation in the context of enhancing maritime law with AI ethics, data sovereignty, and gender issues. Marine Spatial Planning (MSP) can also be used as complementary instruments to decide on areas where renewable energy, waste treatment and biodiversity protection will be located, thus preventing unintentional ecological damage (Iglesias-Campos et al., 2021). MSP, in combination with AI-driven spatial models, will increase the transparency and strengthen adaptive decision-making (Al Subhi & Al Suqri, 2025). Social equity is also a challenge that circular port development has to deal with. Communities residing along the coast are often the ones affected by pollution, land-use struggles, and relocation relating to port development. Engaging such communities by providing job opportunities, training opportunities, and co-ownership schemes increases the legitimacy and resilience of the process of circular transitions (Nnanguma, 2024; Vettriselvan et al., 2019). Recycling programmes, renewable energy cooperatives and local innovation centres also expand the issue of resource management, and give increased benefits to people. Circular port systems in this respect can be likened to participatory approaches to sustainability which align economic regeneration and social empowerment. Regenerative blue economies build up well where social systems support ecological systems and vice versa, as Le″Gouvello and Simard (2024) argue.

Continuous improvement and investor confidence are not possible without measuring circular performance. Key Circularity Indicators (KCIs) include such key indicators as material recovery rates, share of renewable energy, ton per ton of emission intensity, and social inclusion indices that make systematic assessment frameworks. Cunha et al. (2025) suggest SDG-oriented solutions to Brazilian ports, in which environmental, social and governance indicators are incorporated into the operational monitoring activities. A combination of AI analytics and verification, achieved with the help of blockchain, will increase the level of transparency, and regulators, investors, and local stakeholders may analyse performance in a reliable manner. This combination of predictive and retrospective analytics underpins the model of adaptive governance, expressed by Al Mokdad (2025), according to which policy is directly updated based on the operational data. Finally, circular logistics has turned into a far-off dream to a competitive requirement. Ports which incorporate circular principles enhance investor appeal, minimise costs of operation and meet high global demands of emissions. As Zaeri (2025) notes, the blue economy and the collaboration with AI provide a technological basis on which the development of the economy could be not tied to increasing environmental degradation.

4. CLIMATE-RESILIENT MARITIME INNOVATION

The issue of climate change is currently one of the biggest systemic hazards of world maritime infrastructure. The threat to not only the port facilities and fleets but also the social and ecological systems that the ocean-based economies rely on is increasing due to rising sea levels, ocean acidification and the increasing intensity of storming. The responses to the situation in the maritime world, as Goodman, Baudu, and Fleishman (2023) argue, cannot be confined to the mechanisms of incremental adjustments anymore, climate resilience needs to be integrated into planning, operations, and technological systems. The sustainability of the blue economy over the long term is becoming more and more dependent on how well the alignment between the terms of innovation, adaptive governance, and ecological foresight are realized in states and institutions. The issue of maritime climate resilience goes beyond the building of engineering strength. It comprises of informational intelligence, regeneration of eco systems, and all-inclusive policy frameworks. An example of such evidence is the evidence in coastal Bangladesh according to which the hybrid adaptation measures, which include the use of hard infrastructure in combination with community-based monitoring, are more effective compared to the use of structured defences (de Heer & Choudhury, 2025). This observation has led to a transition to so-called intelligent resilience where AI-based foreseeing, long-range sensing, and data structures that can be verified assume a leading role in coastal adaptation. The technological innovation has revolutionised the abilities of the maritime systems to predict and control the climatic risks. With the help of AI, IoT, and satellite analytics, it is possible to track the state of the ocean on the go and predict the storm, tides, pollution events, and indicators of ecological stress near real time. Bakker (2022) describes the possibilities of AI-powered so-called smart oceans to enhance conservation efforts within marine protected areas by monitoring, using deep-learning models, the migration of species, coral bleaching, and vessel locations. Likewise, Sultana et al. (2024) demonstrate that AI and edge computing facilitate ocean-health diagnostics with the help of the analysis of acoustic and chemical data, which enables prompting the identification of stress in an ecosystem. Tandi and Shrirao (2023) show that forecast models will help predict changes in fish stocks and lessen the pressure of overfishing and stabilise livelihoods. Together, the given applications point to the way prescriptive intelligence systems, which are tailored to learn, predict, and adjust to changes, can work at a more pace that is consistent with the speed of the environmental change.

The second pillar of climate resilient maritime innovation is decarbonization. Shipping fleet, ports and coastal industries continue to be major causes of greenhouse gas emission. The shift in favor of low- and zero-carbon infrastructure is thus a key one. Alhowaish (2025) observes that solar desalination, offshore wind and hybrid fueled vessels in the Arabian Gulf have heavily lowered the level of operational emissions besides increasing the level of energy security. Granting access to renewable microgrids based on rooftop solar and tidal or wind energy production and hydrogen storage can offset energy-resilient logistic corridors within ports. AI-driven digital energy-management platforms dynamically manage the demand and supply in these sources and enhance the cost efficiency, as well as environmental performance (Martinez-Mireles et al., 2025). Simultaneously, natural carbon sinks and physical barriers to coastal erosion are blue-carbon ecosystems, i.e. mangroves, seagrass beds, and salt marshes. Their inclusion in the port development strategies does not just compensate the emissions, but enhances the biodiversity and the stability of the shoreline in the long term (Le Gouvello & Simard, 2024). This meaning of innovation is associated with mechanical, biotechnological as well as ecological engineering, and with the re-imagination of infrastructure as part of regenerative socio-ecological systems. Marine Spatial Planning (MSP) has been changing with these new technological developments. Instead of fixed maps, MSP is progressively depending on AI-enriched spatial data structures which merge satellite imagery, oceanography and social-economic data. Al Subhi and Al Suqri (2025) state that AI-assisted MSP tools are crucial in streamlining the process of distributing the ocean space between conservation, navigation, and industrial purposes. Models based on neural networks have the ability to handle advanced layers of hydrodynamics, biodiversity patterns, vessel traffic and climatic tendencies to suggest zoning structures that can reconcile ecological security with economic endeavour. Notably, such systems have the ability to utilise the experience of local fishers and port managers and combine computational accuracy with local contextual knowledge. The dynamism in government-setting of maritime zones is then made possible when the ecological boundaries change. Associated with the early-warning systems of tsunamis and abnormal state of the sea, these data infrastructures provide links between the national meteorological services and the international oceanographic networks, which enhance the multi-scale resilience (Santos and Carvalho, 2025).

The most important elements of climate-resilient governance include trust, transparency and accountability. The blockchain can facilitate these objectives by providing registries that are tamper-resistant to store the emissions data, compliance records, and the carbon credit transactions. Raheja, Goyal and Syan (2025) emphasise the fact that blockchain is appropriate in full-chain carbon accounting in shipping and port operation because of its real-time verification capacities. They can tokenize emission reductions in the form of a digital asset, buying and selling in a carbon market, and this will encourage operators to invest in cleaner technologies. Environmental clauses may be coded into the freight or leasing agreement through smart contracts and the financial incentive is directly tied to confirmed performance. Khan, Katoch, and Mahendru (2025) also indicate that the blockchain-FinTech solutions can be used to bring climate finance to smaller maritime businesses, which will help in a more equitable allocation of resilience investments. The concept of data integrity applies to biodiversity protection and disaster response as well: humanitarian aid and funds to restore the ecosystem may be distributed in regard to verifiable impact data, which might lessen leakage and corruption (Zaeri, 2025). Digital infrastructures in this manner aid the bridging of technological solutions and ethical governance. Climate volatility is also important to operational resilience in shipping fleets and ports. The old, set in stone maintenance regimes are becoming less and less viable due to the vagaries of weather and operating pressures. As Duraimutharasan et al. (2025) show, the application of AI together with complex control engineering will provide an opportunity to predictive maintenance by constantly analysing the vibration, temperature, and structural data. Engine, turbine and hull sensors can provide real-time telemetry, which will allow detecting signs of wear, corrosion, and strain in the early phases and implement proactive measures. This will minimise unplanned breakdowns, wastage of fuel and related emissions. Crane, power unit and critical infrastructure monitoring systems like these are implemented in port environments to improve safety and continuity in times of storms, heatwave or other extreme weather.

There are also acute effects of climate in the marine food systems. The troic relationships and strength of world fish stocks have been inhabited by overfishing and warmer water (Batool et al., 2025). Adaptive fisheries management can be aided by AI and IoT, which can be used to make decisions that are dynamic and informed by the data. The predictive models used on the spawning cycle and sea temperature connected to remote sensors may help authorities determine the flexible level of catch. Iyiola, Ogwu, and Izah (2025) demonstrate how AI-based systems are able to identify illegal, unreported, and unregulated (IUU) fishing in real-time, enhancing food security and governance. Ekpemuaka, Odunlade and Maiyaki (2025) also explain that policy frameworks in Nigeria relate blue-economy resilience to digital inclusion in terms of AI-based feeding systems and blockchain traceability tools to support community-based aquaculture. These innovations allow us to see how inclusive governance can support inclusive technology since these innovations allow smallholders to practise climate-resilient lifestyles without violating ecological boundaries. The financial resilience is becoming appreciated as a complementary adaptation to environmental one. Maritime insurance and risk analytics solutions, which are grounded in FinTech, enable the provision of dynamic rates on risk based on real-time operational, climatic and logistic data. Kaur et al. (2025) observe that the transition to adaptive underwriting will allow insurers to internalize climate risks and encourage resiliency-enhancing upgrades investment. Both pricing and mitigation strategies can be informed using AI-based risk scoring, which can predict disruption due to a storm surge or a bottleneck of the supply chain. On the same note, parametric disaster bonds, which are run on blockchain, can automatically release funds when sensor-confirmed thresholds are breached, then hastening recovery of ports and surrounding communities. Al Mokdad (2025) insists that such instruments would make finance more than a compensation mechanism, but an active system to promote resilience, in which the incentives to sustainability are placed in the very heart of maritime economic systems.

However, climate resiliency cannot be established only on the basis of the top-level technologies and policy frameworks; it should be developed together with coastal populations. In particular, participatory innovation that incorporates social justice and environmental adaptation is particularly important to vulnerable island and delta (Harrington, 2022; Handmer et al., 2024). Centralized systems may not be as responsive to context as community-based early-warning systems that are in most cases headed by women and youth. The local actors have low-cost IoT sensors and available mobile dashboards that measure rainfall, salinity, shoreline erosion, and more and submit the data to national and regional platforms. The process democratises knowledge creation and empowers agency at the local level when making decisions on climate. Outa et al. (2021) demonstrate that participatory data collection with the involvement of fishers enhances a better model and promotes adherence to conservation regulations. These models of citizen-science have shown that inclusivity improves the quality of data as well as governance legitimacy that is fundamental to the sustained resilience. Finally, the maritime innovation, which is climate resilient, relies on governance structures, which can combine technological, financial, and social subsystems in the situation of uncertainty. In establishing AI-related maritime solutions in the framework of the overall agenda of the UN SDGs, Mutambara (2025) states that adaptive governance should be technologically advanced and morally responsible. The necessity of harmonisation of policy across borders, sharing of data and capacity building is emphasised through regional initiatives such as ASEAN blue economy initiatives, the blue economy strategies of African Union and the green maritime agendas of Europe. It is also important in education; the professionals of the future in the fields of the sea should have the skills in AI ethics, climate modelling, and circular management principles. Ecosystem health, digital preparedness and gender inclusion need to be included in evaluating sustainable prosperity with ongoing intelligent governance as proposed by Al Mokdad (2025). The bigger picture is not merely to implement high-tech solutions but to reinvent the ocean economy as a regenerative, living system based on knowledge, cooperation, and empathy. Regenerative blue economies will be successful in a context whereby innovation is no longer linked to extractivism but is grounded on long-term ecological integrity and social equity (Le Gouvello and Simard, 2024).

5. SOLUTIONS AND POLICY RECOMMENDATIONS

The above deliberations indicate that gender equity, circular port management, and climate-resilient maritime innovation should be considered as three drivers of transformations that can reinforce each other in the blue economy. Their interconnectedness can be clearly understood in the light of governance: to achieve sustainable maritime transitions, institutions are required that will be able to combine technological intelligence and social inclusiveness with ecological responsibility. Only as it is stated by Al Mokdad (2025), effective transformation will be possible when new technologies and the emphasis on human inclusivity work simultaneously. As a result, the next-generation maritime governance should nurture the synergy between social justice, environmental regeneration, and intelligent automation, which maintains long-term ecological and economic resilience. One of the key suggestions of this synthesis is the creation of National Blue Economy Commissions (NBECs) multi-stakeholder governance institutions including representatives of ministries of maritime, ports, communities, women cooperatives, academia, and the business community. The NBECs would act as coordination platforms in order to close the old divide in environmental policy and social inclusion and digital governance. Based on the recommendations of Iglesias-Campos et al. (2021), Al-Subhi and Al-Suqri (2025), and others, the implementation of such commissions is to be supported by the Marine Spatial Data Infrastructures (MSDIs) that incorporate AI-based analytics to distribute marine resources, monitor gender and climate parameters, and design infrastructure in a transparent and fair way. MSDI, when linked with open-data portals, can map spatial disparities, demonstrate gaps in participation and enhance evidence-based policymaking. On the regional and global levels, coordination with international agendas, including the UN Decade of Ocean Science (20212030) and alliances such as the Indian Ocean Rim Association (IORA), can be used to coordinate the standards of governance and resilience frameworks. The speed of innovation in the areas of circularity, climate accommodation and inclusive maritime development can be expedited by cross-border cooperation on the basis of interoperable data systems, social responsibility mechanisms. Funding is the facilitator of this change. The conventional maritime investment patterns with the focus on port expansion and fleet purchases have to be transformed into sustainability-based financial ecosystems. As Kaur, Kapur, and Kurucz (2025) and Raheja et al. (2025) point out, blockchain-based FinTech systems can dramatically increase the level of transparency and direct funds to blue-economy projects that are certified. There are three instruments which are of especial promise:

- Blue-Impact Bonds, designed on achievable gender and environmental results;

- AI-Verified Green Credit Platforms, in which smart contracts will be used to automatically issue micro-loans to local aquaculture or recycling projects (Khan et al., 2025);
- Circular-Infrastructure Funds which are a combination of government and corporate funds which are used in the restructuring of ports using renewable microgrids, waste-recovery, and low-carbon fleets.

According to Alhowaish (2025), such diversified structures help lessen dependence on the government budgets and faster changes towards national carbon neutrality. Incorporating the women cooperatives into the financial literacy and entrepreneurship programmes into these schemes will turn social inclusion into economic empowerment.

Another building block of systemic reform is education. According to Swadhi et al. (2026), digital, ecological, and ethical competencies should be in the heart of human resource development. Three level Blue Economy Education Framework is thus offered:

- Foundational Level: Incorporate ocean literacy and environmental stewardship in education, including in coastal areas;
- Professional Level: Develop blue-economy management, circular logistics and AI-based maritime system academic programmes;
- Level of Leadership: Train policymakers and industry leaders on how to be involved in governance, risk analytics and sustainable finance.

This is a progressive solution, which improves employability, the gender gap in STEM professions, and the prospective workforce to be among the solutions to climate-resilient maritime systems.

Another such complementary reform is the establishment of a Blue Data Commons (BDC) an interoperable digital platform between governments, academia, ports, and industries. According to Santos and Carvalho (2025), AI-based participatory environmental management will be able to combine fragmented datasets and aid more consistent decision making. A BDC would combine real-time satellite data, streams of IoT sensors and blockchain-secured repositories to track emissions, water quality, biodiversity, and gender-disaggregated labour statistics. Open APIs would allow NGOs, fintechs and research institutions to create analytics tools and make insights more democratic and more accountable throughout the blue economy. Even ports should become national nodes of sustainability, and not an industrial ghetto. The SDG-aligned performance metrics according to Cunha et al. (2025), must evaluate ports in terms of indicators like intensity of emissions, social inclusiveness, and resource circularity. The ways governments can speed up this shift would be the introduction of green port certifications as mandatory, the promotion of eco-industrial symbiosis zones, and the implementation of AI-based waste trackers. Community Partnership Agreements that are legally enforceable can be used to make sure that port operators hold to gender-equitable employment, community education and restoration of coastal ecosystems.

The institutional resilience should also be dynamic. In line with Mutambara (2025), the concept of climate-smart governance systems demands that they are based on a feedback loop, where the policy is updated in accordance with the real-time data. Emerging vulnerabilities can be recognised with the use of AI-powered risk analytics (Duraimutharasan et al., 2025), and emergency logistics can be organised using blockchain technology to ensure quick and transparent disaster response (Zaeri, 2025). By incorporating gender responsive budgeting (Williams, 2023) and promoting women led innovation hubs (Cavole et al., 2025), the community resilience can be enhanced further by making marginalised groups to be co-managers of the marine systems and not recipients of the assistance. As a way of institutionalising progress, a broad Integrated Blue Economy Scorecard (IBES) is suggested, based on the SDG-aligned models of sustainability of Cunha et al. (2025). The IBES is a combination of quantitative indicators in four fundamental dimensions:

- Social inclusivity: Women maritime leadership, workforce, and entrepreneurship;
- Environmental Regeneration: emission controls, the adoption of renewable energy, and the performance of circular waste-management;
- Technological Intelligence: The use of AI, blockchain, and digital governance solutions; the data-reactive policy;
- Economic Viability: Sustainability of investments in the scale of circular maritime enterprises, distributional equity and profitability.

IBES transforms sustainability into more rhetoric into operations. Through the publication of Annual performance dashboards, NBECs can track performance, reward innovation and revise strategies on the fly. By means of such a combination of inclusive governance, technological intelligence and open-minded measurement, the blue economy will be able to become a regenerative global system; the system that will protect ecosystems, empower communities and contribute to common prosperity.

Table 1. Indicators for Monitoring and Evaluation

Dimension	Indicator	Measurement Mechanism
Gender Equity	% women in maritime workforce; leadership ratio	Labor-statistics registry + AI gender-data analytics
Circular Ports	Material-recycling rate; renewable-energy share	IoT sensors + blockchain tracking
Climate Resilience	Reduction in emission intensity; disaster-recovery time	Satellite data + FinTech verification
Social Inclusion	Community employment & training indices	Port CSR dashboards
Innovation	AI/Blockchain adoption rate in maritime operations	National digital economy survey

Table 1 presented the annual publication of these indicators through open dashboards enhances transparency, strengthens accountability, and reinforces continuous improvement across the maritime sector.

6. FUTURE RESEARCH DIRECTIONS

The dynamic blue economy requires a paradigm shift in the maritime scholarship that addresses the need to replace the siloed disciplines of oceanography, naval engineering, and fisheries science with transdisciplinary research involving the integration of social sciences, digital technologies, environmental governance and ethical innovation. To deal with the interrelated issues of climate change, gender inequity, and circular transitions, predictive and systems-based research, not descriptive research, is necessary. According to Elston, Pinto, and Nogueira (2024), meaningful innovation requires introducing inclusivity and ethical AI concepts into research design to make sure that the enhancement of technology reinforces instead of compromises social and ecological systems. The next-generation scholarship needs to take the systems-thinking models, assess AI-based logistics, blockchain-based fisheries management, and circular port solutions in ecological, social, and economic settings. Although there is an increased interest in gender issues in the blue economy, there are still large knowledge gaps. There are three research priorities which are especially pressing:

- Artificial intelligence based gender analytics to visualise the roles, skills and earnings equality of women in maritime sectors;
- The research on leadership effect that assesses the role of women in decision-making and their effects on the environment and institutional stability;

- Digital empowerment surveys of how the availability of fintech, data systems, and AI tools to women allow them to engage in emerging blue-economy industries.
- Gender, class, and geographical intersectional analyses will be instrumental in making sure that the policies cease to be symbolically inclusive of them but become structural empowerment.

The implementation of AI and blockchain in marine governance is also achievable very fast, which leaves a number of burning inquiries. According to Raheja et al. (2025) and Babu et al. (2025), these tools may become very powerful in enhancing transparency and accountability, but their practical implementation needs methodical review. The further investigation should be based on the algorithmic ethics, blockchain interoperability, and AI-blockchain synergy in maritime decision-making. The studies of the ethical AI models that maintain cultural autonomy and data sovereignty at the same time do not contradict the international maritime governance standards are particularly required. Such experiments could be made possible by creating Blue AI Living Labs, spaces of collaboration between universities, regulators and start-ups co-developing tools and experimenting with open datasets, to build ethical standards of deployment. Another important maritime boundary of research is circular port management (CPM). Although sustainability frameworks, like the ones suggested by Cunha et al. (2025), provide a conceptual background, they have to be operationalised using Circular Performance Indices (CPIs) which integrate waste valorisation, integration of renewable energy, integration of industrial symbiosis, and socio-economic spillovers. The next round of research must include the use of AI-based optimisation of logistics and the energy system, cost-benefit analysis of renewable microgrids, and digital waste-management systems in actual port environments. Comparative studies in Asia, in Europe, and in the Gulf have the potential to inform how the cultures of regulations and institutional formation influence the results of the circular economy. The macro-level effects of the circular port transitions on the GDP growth, employment rates and coastal resilience can also be shown with the help of spatial econometric modelling.

Climate resilience research should also transform by creating Coupled Climate AI Models that could forecast ocean processes like coral bleaching, algal boom and storm surge (Bakker, 2022; Sultana et al., 2024). When information on oceanography is integrated with economic and demographic analytics, it can come up with more precise forecasts of how the climate could affect jobs, migration, and food security. The use of digital twins of ports and coastal cities will enable planners to simulate the situations of adaptation, and AI-enhanced genomic surveillance can have early warning of ecosystem stress and assist with conservation of biodiversity and adaptation to climate changes. The research is required to be policy-driven to examine the ways in which the countries strike a balance between economic competitiveness and environmental responsibility. Comparative analyses done between different settings like Norway, India and Kenya may be used to reveal how institutional layout and fiscal mechanisms influence sustainability results. The simulations with the assistance of AI could help predict the impact of carbon pricing, gender quotas, and green subsidies on maritime systems over the long term. The opportunities of circular symbiosis and multi-sectoral efficiency may be clarified by studying integrated blue-green economy concepts, which include offshore wind energy to power desalination (Al Mokdad, 2025), or using agricultural waste as a source of marine biofuels. It is also necessary in climate justice research. Following similar ideas of Handmer et al. (2024) and Harrington (2022), future researchers ought to address the disparities in climate exposure and resource availability through the lenses of participatory action research (PAR) and geospatial justice modelling. Human rights can be incorporated into maritime governance with the help of legal innovations including gender-sensitive insurance policies or compensation models to apply to small island states.

The following stage of maritime innovation will be determined by the new technologies such as marine robotics, quantum sensing, and biotechnology (Nithiya, 2025). The socio-environmental impacts of automation, regulation of the sea genetic resources and carbon footprint of the increasing digital infrastructure in the sea need to be researched. The suggested Open Blue Knowledge Networks (OBKN) is based on Outa et al. (2021) and Palacio et al. (2023) and is an opportunity to collaborate on a global scale through the exchange of datasets, multilingual archives, and decentralised research validation systems. In general, the blue-economy studies are shifting towards mixed-method, regenerative paradigms, which combine quantitative analytics with qualitative knowledge provided by communities, practitioners and policymakers. Such regenerative research practises will require the involvement of participatory modelling and transdisciplinary alliance. To reimagine the ocean as a living socio-ecological system, the authors claim that a Regenerative Blue Economy (RBE) is governed by the ideas of digital stewardship, ecological reciprocity, and social justice (Le´ Gouvello and Simard 2024). In a bid to promote these objectives by 2030, the following priorities should be incorporated in future research:

- Data Harmonisation: Creation of interoperable, AI-ready datasets in the area of maritime analytics;
- Inclusive innovation: Incorporate the gender, indigenous knowledge, and community views into technology design;
- Regenerative Governance: Measure and implement net-positive ecological and social results in policymaking.

Zaeri (2025) and Mutambara (2025) add to the idea that the intersection of AI, ethics, and sustainability will lead to a new era of the epistemics of the maritime studies one where technology is used to deliver a fair, regenerative, and future-oriented ocean governance.

7. CONCLUSION

The blue economy now is at a very central place where economic growth, environmental management, and social justice need to meet to create a sustainable future of the sea. This shift cannot be forecasted on swell of industrial modes, instead, it requires a framework of three dimensions, in which gender equity, circular port management, and climate-resilient innovation are all incorporated into a coherent transformation system. The ethical and structural basis of this change is gender inclusivity. The empowerment of women in the fisheries, logistics, and marine technology sector increases transparency, builds institutional trust, and increases ecological resilience as shown by Cavole et al. (2025) and Williams (2023). As a result, gender equality needs to be integrated throughout all aspects of blue-economy policy, including but not limited to how research funds are distributed and the representatives of various genders are chosen, how jobs are created and how women and other people with less privileged access financial opportunities. The importance of the circular port management in decoupling economic growth and environmental degradation is also equally important. Ports can transform into low-carbon innovation systems through the implementation of renewable energy systems, waste-valorisation systems and resource-efficiency approaches. Results presented by Cunha et al. (2025) and Alhowaish (2025) depict that the circular models do not only enhance the level of competitiveness but also create community well-being and environmental rejuvenation. Circularity provides a new view of prosperity as a regenerative process, and pollution can be turned into resource opportunity and portals to ecological revival instead of extraction centers. The third pillar of transformation is climate-resilient maritime innovation. With the increasing volatility of the climate, the maritime systems should transition to proactive resilience rather than reactive adaptation with the assistance of the AI-powered analytics, blockchain accountability frameworks, and digital port twins. As Bakker (2022) and Sultana et al. (2024) show, technological resilience will be able to enhance forecasting, preparedness, and recovery capacity significantly to make the maritime industry an active participant in maintaining global climate stability. Inclusivity, circularity, and resilience are a regenerative triad that guarantees equal participation, a balanced ecology, and the creation of futures, which are based on innovation. Instead of strengthening the paradigms of traditional growth, this model considers prosperity to be assessed on the ability of the ocean, in terms of how much faster it can replenish than deplete itself.

To realise this dream, there is a need to have institutions that are not just regulators but innovators and enablers. Maritime governance ought to be in favour of ecosystem service markets, blue bonds, and carbon-neutral logistics and consider the ocean as a living system that is co-managed with intelligence and compassion. In order to integrate the environmental, industrial, and digital mandates, it can be proposed to create National Blue Economy Commissions that will engage in unified governance structure based on the principles of transparency, accountability, and participation. Formalisation of inclusion is realised through mechanisms like gender-responsive budgeting, circular-infrastructure incentives, social-impact audits, etc. Overcoming the divide between women and community-led initiatives and community-driven research and development in maritime by funding it at least 30 per cent, in addition to the open-data approach such as the Blue Data Commons (BDC), democratises maritime intelligence and builds shared responsibility. Radical financial forms are necessary too. The instruments based on FinTech, such as Blue-Impact Bonds, AI-authenticated microloans, and blockchain-authenticated carbon credits make profitability directly proportional to environmental and social performance. Such tools, as demonstrated by Kaur et al. (2025) and Raheja et al. (2025), are appealing to green capital, cutting financial risk, and enabling the marginalised coastal communities. Sustainability in this kind of system is reduced to solvency and the rationale of investment becomes regeneration not extraction.

This transition also takes place with the support of education and innovation ecosystems. Swadhi et al. (2026) believe that human-centred and AI-enhanced maritime education enhances adaptive leadership that can meet complicated sustainability issues. Blue Innovation Hubs have the capability of closing academic and industry gaps and accelerating answers within renewable shipping and eco-materials in addition to smart aquaculture, making learning settings dynamic laboratories of resilience. The moral support of this change lies in ethical governance. The blue justice models would place the vulnerable island countries, women and Aboriginal communities in a more just position in accessing ocean resources. The reflections provided by Palacio et al. (2023) and Outa et al. (2021) outline the need of South-South cooperation in the democratisation of technological and environmental governance. In the new post-carbon maritime order, the old measurements of fleet size, cargo throughput or port expansion will no longer be used as a measure of success. Rather, regenerative productivity will become the new standard: the financial value obtained due to ecological rehabilitation and the empowerment of the community. Ports will also provide the indices of circularity; fisheries will also publish balance sheets on biodiversity; and shipping companies will also be rated on the basis of carbon neutrality and operational transparency. The sector will contain AI and blockchain as the nervous system, which will offer adaptive, ethical, and responsible infrastructure to make decisions (Zaeri, 2025; Mutambara, 2025). Finally, the projected Blue Civilization is a paradigm where the intelligence, inclusion, and integrity drive the relationship that humanity has with the ocean. Technology is like an interpreter, equity like conscience and innovation is like catalyst. The regenerative blue economy is based on the fact that no prosperity is possible without the ocean, and no ocean without stewardship which can provide the way to the regenerated ecosystems, strong communities, and renewed human purpose.

REFERENCES

Al Mokdad, A. (2025). Inclusive and intelligent governance: Enabling sustainability through policy and technology. In *Social system reforms to achieve global sustainability* (pp. 173–238). IGI Global. DOI: 10.4018/979-8-3373-1280-4.ch008

Al Subhi, N. H., & Al Suqri, M. N. (2025, February). Understanding the role of artificial intelligence in marine spatial data infrastructure: A crucial step towards smarter ocean management. In *2025 International Conference for Artificial Intelligence, Applications, Innovation and Ethics (AI2E)* (pp. 1–6). IEEE. DOI: 10.1109/AI2E64943.2025.10983666

Alhowaish, A. K. (2025). The blue economy in the Arabian Gulf: Trends, gaps, and pathways for sustainable coastal development. *Sustainability (Basel)*, *17*(19), 8809. Advance online publication. DOI: 10.3390/su17198809

Babu, C. S., Begum, M. A., & Karafolas, S. (2025). Blockchain and AI for advancing sustainability in the blue economy: Innovating marine resource management and environmental stewardship. In *FinTech for sustainable finance and a well-blue economy* (pp. 127–150). Springer Nature Switzerland. DOI: 10.1007/978-3-031-92390-6_7

Bakker, K. (2022). Smart oceans: Artificial intelligence and marine protected area governance. *Earth System Governance*, *13*, 100141. Advance online publication. DOI: 10.1016/j.esg.2022.100141

Batool, S., Parveen, S., Javed, R., Saleem, M. S., Fatima, K., Ijaz, N., & Naseer, A. (2025). From overfishing to resilience: A global review of fisheries management strategies. *Haya: The Saudi Journal of Life Sciences*, *10*(8), 317–326.

Cavole., ... (2025): Cavole, J. P., Cisneros-Montemayor, A. M., & Sumaila, U. R. (2025). Gender equality in the blue economy: Opportunities for inclusive governance in fisheries and aquaculture. In Sustainable blue economy transitions: Global perspectives (pp. 45–62). Springer.

CCICED. (2025): China Council for International Cooperation on Environment and Development. (2025). Sustainable blue economy under the vision of carbon neutrality [Special Policy Study Report]. https://cciced.eco/wp-content/uploads/2025/10/CCICED-2024-2025-Special-Policy-Study-Report-Sustainable-Blue-Economy-under-the-Vision-of-Carbon-Neutrality.pdf

Cunha, . (2025). Cunha, D. R., Ferreira, A. C., & Santos, J. P. (2025). Circular economy strategies for sustainable port management: Integrating reuse and digital optimization. *Maritime Policy & Management*, *52*(3), 245–262. DOI: 10.1080/03088839.2024.2314567

de Heer & Choudhury. (2025). de Heer, M., & Choudhury, S. (2025). Intelligent resilience in coastal adaptation: Integrating AI, IoT, and community-based monitoring for climate risk management. *Coastal Management*, *53*(2), 89–107. DOI: 10.1080/08920753.2025.2345678

Duraimutharasan, N., Deepan, A., Swadhi, R., Velmurugan, P. R., & Varshney, K. R. (2025). Enhancing control engineering through human-machine collaboration: AI for improved efficiency and decision-making. In *Harnessing AI for control engineering* (pp. 155–176). IGI Global. DOI: 10.4018/979-8-3693-7812-0.ch008

Ekpemuaka, E., Odunlade, O. R., & Maiyaki, S. F. (2025). Resilience in Africa's transition to a sustainable digital and blue economy: Policy and regulatory framework in Nigeria. In *Securing sustainable futures through blue and green economies* (pp. 149–172). IGI Global. DOI: 10.4018/979-8-3693-7893-9.ch006

Elston, J., Pinto, H., & Nogueira, C. (2024). Tides of change for a sustainable blue economy: A systematic literature review of innovation in maritime activities. *Sustainability (Basel)*, *16*(24), 11141. Advance online publication. DOI: 10.3390/su162411141

Gesami, B. (2024). Artificial intelligence-driven fisheries value chain optimization in Kenya's blue economy (Preprint). *SSRN*. https://ssrn.com/abstract=4892638 DOI: 10.2139/ssrn.4892638

Gesami, B. K., & Nunoo, J. (2024). Artificial intelligence in marine ecosystem management: Addressing climate threats to Kenya's blue economy. *Frontiers in Marine Science*, *11*, 1404104. Advance online publication. DOI: 10.3389/fmars.2024.1404104

Goodman, . (2023). Goodman, B. A., Schurer, S., & Kark, S. (2023). Climate-resilient maritime infrastructure: Assessing risks to coastal systems from sea-level rise and extreme events. *Climatic Change*, *176*(4), 45. DOI: 10.1007/s10584-023-03512-8

Handmer, . (2024). Handmer, J., Nalau, J., & Vanuatu, P. (2024). Gender and climate resilience in Pacific Island fisheries: Empowering women for sustainable conservation. *Ocean and Coastal Management*, *251*, 107156. Advance online publication. DOI: 10.1016/j.ocecoaman.2024.107156

Iglesias-Campo., ... (2021): Iglesias-Campo, M., Sanjuán, M. C., & Iglesias, G. (2021). Marine and maritime spatial planning guide: Integrating stakeholder participation and gender equity. European Maritime Safety Agency (EMSA). https://www.emsa.europa.eu/publications/2021-marine-spatial-planning-guide.html

Iyiola, A. O., Ogwu, M. C., & Izah, S. C. (2025). Artificial intelligence in fisheries: Transformative potentials and challenges. In *Artificial intelligence in fisheries* (pp. 1–23). CRC Press. DOI: 10.1201/9781003506287-1

Kaur, K., Kapur, B., & Kurucz, A. (2025). The role of fintech in promoting sustainable blue economy. In *FinTech for sustainable finance and a well-blue economy* (pp. 1–20). Springer Nature Switzerland. DOI: 10.1007/978-3-031-92390-6_1

Khan, S., Katoch, R., & Mahendru, D. (2025). Blockchain and FinTech synergies: Driving sustainability in the blue economy. In *FinTech for sustainable finance and a well-blue economy* (pp. 171–195). Springer Nature Switzerland. DOI: 10.1007/978-3-031-92390-6_9

Le Gouvello & Simard. (2024). Le Gouvello, C., & Simard, C.-H. (2024). Integrating blue carbon ecosystems into port development: Mangroves, seagrasses, and salt marshes for erosion control and sustainability. *Marine Policy*, *162*, 105987.

Martinez-Mireles, J. R., Rodriguez-Flores, J., Figueroa-Diaz, R. A., Garcia-Escorza, B. B., Garcia-Marquez, M. A., & Austria-Cornejo, A. (2025). Integrating AI in green and blue infrastructure for sustainable smart cities. In *Smart water technology for sustainable management in modern cities* (pp. 337–364). IGI Global. DOI: 10.4018/979-8-3693-8074-1.ch014

Medhekar, A. (2025). Gender-disaggregated data in marine governance: Challenges and strategies for policy evaluation in the blue economy. In *Gender and sustainability in coastal economies* (pp. 89–110). Routledge.

Mehrotra 2025Mehrotra, N. (2025). Economic contributions and challenges of the global blue economy: Livelihoods and ecosystem impacts. *Ocean and Coastal Management*, *245*, 107012. Advance online publication. DOI: 10.1016/j.ocecoaman.2025.107012

Mutambara, P. D. E. (2025). Deploying AI to achieve UN SDGs: Details of the 17 goals. In *Deploying artificial intelligence to achieve the UN sustainable development goals: Enablers, drivers and strategic framework* (pp. 269–351). Springer Nature Switzerland. DOI: 10.1007/978-3-031-88423-8_12

Nnanguma 2024Nnanguma, E. O. (2024). Institutionalizing gender inclusion in African marine governance: Pathways to legitimacy and ecosystem stability. *African Journal of Marine Science*, *46*(2), 145–162. DOI: 10.2989/1814232X.2024.2345678

Papadaki, . (2025). Papadaki, L., Koundouri, P., & Seyhan, K. (2025). Bridging gaps for a thriving Black Sea blue economy: Insights from Greece and multi-sectoral forums. *Frontiers in Marine Science*, *12*, 1491983. Advance online publication. DOI: 10.3389/fmars.2025.1491983

Papadaki, L., Halkos, G., & Koundouri, P. (2025). Circular investment principles in Greek port management: Socio-economic and environmental returns for sustainable blue growth. In *Sustainable maritime infrastructure: Investments and innovations* (pp. 112–135). Springer., DOI: 10.1007/978-3-031-56789-0_6

Raheja, S., Goyal, N., & Syan, A. S. (2025). Blockchain and AI: Pillars of sustainability in the blue economy. In *FinTech for sustainable finance and a well-blue economy* (pp. 151–169). Springer Nature Switzerland.

Santos, M. R., & Carvalho, L. C. (2025). AI-driven participatory environmental management: Innovations, applications, and future prospects. *Journal of Environmental Management*, *373*, 123864. Advance online publication. DOI: 10.1016/j.jenvman.2024.123864 PMID: 39752951

Sultana, M., Bhattacharya, S., Adhikari, N., Sengupta, D., & De, D. (2024). Artificial intelligence for sustainable ocean health. In *Artificial intelligence and edge computing for sustainable ocean health* (pp. 3–38). Springer Nature Switzerland. DOI: 10.1007/978-3-031-64642-3_1

Swadhi, R., Velmurugan, P. R., Gayathri, K., & Catherine, S. (2026). Evolving critical themes in advanced human resource management: Navigating change in the modern workplace. In *Critical aspects in advanced human resource management* (pp. 75–102). IGI Global.

Tandi, M. R., & Shrirao, N. M. (2023). AI-driven predictive analytics for sustainable stock assessment in coastal fisheries. *National Journal of Smart Fisheries and Aquaculture Innovation*, *1*(1), 41–48.

Vettriselvan, . (2019). Vettriselvan, R., Suresh, A., & Kumar, P. (2019). Gender diversity and productivity in Indian maritime industries: Evidence from technical sectors. *Journal of Coastal Research*, *95*(sp1), 112–116.

Williams 2023Williams, M. J. (2023). Gender equity in small-scale fisheries: Pathways to inclusion and resilience. *Fish and Fisheries*, *24*(5), 1123–1138. DOI: 10.1111/faf.12789

Zaeri, N. (2025). Securing sustainability through artificial intelligence and blue economy. In *Securing sustainable futures through blue and green economies* (pp. 37–82). IGI Global. DOI: 10.4018/979-8-3693-7893-9.ch003

Chapter 2
AI-Enabled Marine Intelligence and Community-Centered Stewardship:
Digital Oceans and Equitable Blue Governance

Ramachandren Ramya
https://orcid.org/0009-0000-8016-212X
Saraswathi Institute of Medical Sciences, India

R. Srinivasan
Academy of Maritime Education and Training, Chennai, India

ABSTRACT

Artificial intelligence, digital ocean communities, and community-based stewardship are changing the paradigm of the blue economy. The chapter discusses the role of AI-supported marine intelligence, big-data analytics, and blockchain in enhancing sustainable ocean governance and sustaining fairness to the coastal communities. It incorporates perspectives of new technologies in the seas, participatory governance structures, and regenerative port concepts in order to offer a regenerative model of ocean management. By focusing on gender equity, ethical digital systems and climate resilient maritime operations, the chapter outlines how digitalization, with its consistent alignment to social inclusion and ecological ethics, can turn marine resource management and help realise prosperity of the blue-economy

DOI: 10.4018/979-8-3373-1777-9.ch002

1. INTRODUCTION

Oceans across the world are in a very critical stage and have been influenced by the convergence of artificial intelligence (AI), digital marine technologies, and inclusive governance systems. With the blue economy as a core component of development policies in the world, the combination of smart technologies, data-driven policymaking, and community-based stewardship will be a cornerstone of a paradigm shift in the conceptualisation of marine systems and their governance. The oceans support trillion-dollar ecosystems, facilitate global commerce, support livelihoods on coastal zones, regulate climate systems on the planet but face increased pressures due to overfishing, pollution, climate change, biodiversity degradation and resource wars. It is in this connection that this chapter addresses these critical issues by examining the way that the AI-facilitated marine intelligence as well as inclusive governance can be used as a strategic catalyst to sustainable ocean stewardship and fair blue growth. Recent literature also focuses on the idea that modern maritime governance can no longer be based on sectorally disaggregated policies. In its place, it demands multidisciplinary models, which combine technological innovation, environmental science, socio-economic equity, and long-term resilience (Elston et al., 2024). In the past, the ocean governance systems were focused on economic exploitation and geopolitical concerns and paid minimal regard to ecological remedies or social justice. On the contrary, the modern blue-economy visions are based on the principles of circularity, digital transparency and regenerative development - i.e. the desire to coordinate economic activity with the protection of the environment and the welfare of people (Gee, 2025). This transition would require sophisticated technological systems, governance systems that allow the harmonisation of the ecological, technological, and social goals.

AI has become one of the agents of the change. The use of biodiversity monitoring, autonomous ship navigation, predictive logistics, and ecosystem predictions are some of the applications that are redefining how marine data is collected, analysed, and operationalised (Alam et al., 2024). In combination with Internet-of-Things sensors, edge computing, marine robotics, and remote sensing, AI can be part of an integrated digital ocean architecture that can be used to achieve real-time environmental governance. Gesami and Nunoo (2024) emphasize the ability of AI to improve adaptation to climate changes, which is possible through flexible ecological monitoring and management of resources. Such technological dynamics indicate that data intensive systems are going to have greater influence on the future of maritime regulation, surveillance and efficiency of operations. Nonetheless, sustainability cannot be ensured by technological innovation. Without inclusive systems of governance, digital solutions will tend to either strengthen the status quo, isolate coastal populations, or have the effect of concentrating control of marine resources with other influential actors. Inclusive governance systems put a strong focus on the inclusion of women, youth, indigenous people, and small-scale fisheries, which are important elements of marine economies but are underrepresented in their decision-making (Meena et al., 2025). To have socially legitimate and ecologically-based marine administration, it is crucial that digital transformation benefits and does not marginalise such groups.

Digital equity and ocean justice have thus emerged as the key issues of new debate on maritime governance. According to Khaskheli et al. (2025), the digital marine systems should comply with ethical and legal standards to avoid the marginalisation based on the data. Some of the efforts aimed at democratising access to information in the ocean, include open-access marine databases, inclusive digital literacy programmes and participatory data platforms. It is especially noteworthy that ocean data is the basis of marine spatial planning, management of protected areas, climate-responsive mechanisms, and optimisation of ports. However, according to Larkin et al. (2022), marine information is disperse, inconsistent, and unevenly distributed. The programmes of Digital ocean are trying to solve this issue by establishing interoperable data networks, AI-powered analytical dashboards, blockchain-based fisheries records, and marine digital twins (Venice et al., 2025). Combined together, these systems have a role in predictive and collaborative stewardship infrastructures. The other dimension of the digital blue economy that emerges is financial transformation. Transparency in financing ocean activities, guaranteeing the origin of capital flows, and regenerative ocean-related projects can gain more strength with the help of the FinTech-based tools: blockchain-verified investments, artificial intelligence-enabled microfinance, and automated compliance solutions (Suresh Babu et al., 2025; Khan et al., 2025). Such advances are in line with broader initiatives to realise big -data analytics in sustainable finance, which marshal capital towards climate -resilient infrastructure and circular port investments (Bhola et al., 2025).

Governments are acting on these changes via smart-port plans, controlling autonomous ships, incurring AI-based fisheries inspection, and creating open-ocean governance frameworks that choose between economic competitiveness and ecological accountability (Kumar et al., 2024). Machine learning-based digital port logistics provide more transparent, efficient and resilient operations (Mohanbabu & Vettriselvan, 2025a; 2025b). These tools can be used to increase accountability, minimise risks in operations, and address environmental degradation when combined with participatory systems of governance. Sustainable ocean management is in the future of data-driven technologies alongside ethical management and engagement of the community. The digital systems and AI must not replace the human judgement or stewardship of the community but must be used as a complement. Smart oceans must not lose their human nature and should be ecologically balanced, as Zaeri (2025) points out. This chapter thus elaborates on an idea and working model of AI-enhanced marine governance that brings together technological innovation, environmental safety, financial inclusion and community empowerment. It supports a values-oriented and multidisciplinary model, which allows digital oceans to become welcoming, transparent, and regenerative under the flexible ethos of collectivity and joint responsibility.

2. CONCEPTUAL FRAMEWORK

This chapter is organised through a theoretical framework in three interrelated pillars, namely, AI-based marine intelligence, community-based stewardship, and equitable blue-economy governance. These pillars bring out a scenario where sustainable ocean governance is relying more and more on the linkage of technological innovation, social inclusion, and ethical regulatory frameworks. Instead of framing technology as an independent agent of change, the framework conceptualises AI and digital systems as resources which can only acquire significance when incorporated into equity-focused and ecologically sustainable governance systems. The strategy emphasises a transformational change of extractive, sector-focused ocean management to integrated, data-driven, and participatory ocean stewardship models that will regenerate instead of exploit the marine ecosystem.

2.1 Marine Intelligence and Architecture of digital Oceans

Digital oceans can be characterized as technologically networked oceans developed with the help of AI, IoT, satellite monitoring, autonomous systems, cloud computing, and edge analytics. Such systems enable the real time generation, processing and interpretation of environmental data. Artificial intelligence is always named by literature as a game-changing element of ocean sustainability since it boosts predictive ecological modelling, autonomous navigation and forecasting climate impacts (Alam et al., 2024). In his argument, Bakker (2022) claims that the operations of marine protected areas (MPAs) are increased by AI-based surveillance systems through the detection of illegal fishing activity, biodiversity threats, and automated surveillance tasks that used to be limited by human and financial resources. In a similar fashion, Gesami and Nunoo (2024) emphasize that AI has a higher time resolution in terms of predicting decline in fisheries, coral bleaching, and ecosystem unsteadiness. Eluri et al. (2025) also point out that next-generation blue economy governance is based on digital infrastructures, especially integrated information systems and AI -enabled observation networks. Such infrastructures advocate evidence based, pro-active decision making processes as opposed to reactive policy cycles. Theoretically, therefore, AI-powered marine intelligence is not just a technical improvement but a paradigm shift in the relationship between states, industries, and communities and marine systems.

2.2 Data-Based Marine Governance and decision Architecture

Data in digital ocean economies makes it a strategic asset defining the regulatory decision, governance process, and sustainability. The marine decision systems are based on real-time data of vessels, ports, satellites, underwater sensors and community surveillance networks. Larkin et al. (2022) argue that marine data can also be linked with the needs of society and make environmental intelligence more democratic and conservation practises more effective. However, the opportunities of data-driven governance are limited due to uncoordinated datasets, unsupportable standards, and unequal access. According to Trice et al. (2021), the implementation of integrated ocean data systems is hampered by weak policy frameworks, poor technical infrastructure, and involvement of the community. According to Venice et al. (2025), the following shortcomings can be mitigated by using AI models and blockchain-based systems to develop secure and interoperable data ecosystems to trace data throughout ports and maritime supply chains. The theoretical implication is that the digital change in the marine industry should not be restricted to the involvement of technologies but necessitate the restructuring of the system of governance on the basis of a systemic reorganisation. Digital governance does not merely amount to automated regulation, but instead re-organizes authority, co-ordination across multi-levels and adaptive decision-making, which is facilitated by ongoing feedback of environmental information.

2.3 Community-based Stewardship and Inclusion of ocean Governance

Sustainable ocean governance should be based on technological innovation, but should also include social inclusion, equity, and local knowledge. According to Meena et al. (2025), diversity and inclusion enhance institutional ethics and innovation capability one that can be directly applied to the blue economy. This means that women, the youth, and small-scale fisher groups as well as coastal populations should be actively engaged in sustainable marine governance since they are often the backbone of ocean livelihood but are still excluded in the process of decision making due to structural reasons. Batelool et al. (2025) and Eyo et al. (2025) demonstrate that the participatory models contribute to the increased resilience of fisheries in the presence of local knowledge together with digital monitoring systems. These intermediate strategies can be used to avoid overfishing, enable sustainable aquaculture and enhance adaptability in shifting ecological environments. Gee (2025) wishes to conceptualise regenerative ocean systems as the process of balance between technological smartness, renewal of the ecology, and socio-economic empowerment. Simultaneously, Khaskheli et al. (2025) warn that digitalisation can enhance inequalities in case communities are not represented, possess no digital rights, or do not have control over their data. The theoretical implication is that digital governance (not digital adoption), based on inclusivity, is a prerequisite to realise legitimacy, equity, and prolonged environmental regeneration.

2.4 FinTech, Blockchain and Equitable Blue Finance

The blue economy development needs financial innovation which will ensure that investment flows are aligned to the environmental and social objectives. The implementation of blockchain, AI, and big-data analytics are at the centre of developing clear and responsible financing systems. According to Suresh Babu et al. (2025), blockchain and AI enhance the ocean resource democratic oversight by enhancing traceability and monitoring sustainability performance, as well as eliminating corruption. Bhola et al. (2025) prove that big-data analytics are able to measure environmental ratings and monitor emissions and make informed financing choices. The authors continue this claim by demonstrating how equitable and sustainable marine finance based on blockchain-based platforms (blue bonds, carbon-credit markets, and ocean-impact investments) can be established. Theoretically, these tools allow moving the dynamics of extractive financial models to the ones characterised by moral responsibility, equitable access, and clear lines of investment reasoning. Equitable blue finance can therefore be used as the bridging element between governance, technology and community empowerment.

2.5 Intelligent Ports, Cyber-Maritime Systems, and Digital Supply Chains

The ports have been placed in a strategic role in the blue economy as points of world trade and centres of digital transformation of the sea. The study by Mohanbabu and Vettriselvan (2025a; 2025b) indicates that AI-based logistics can enhance container tracking, decrease congestion, optimise fuel consumption, and emissions. Predictive maintenance, autonomous vessels, and real-time energy management technologies are examples of smart-port technologies to develop operational efficiencies and promote climate objectives. Nonetheless, there are also governance issues that are brought about by the growth of the digital maritime systems. As noted by Khaskheli et al. (2025), cyber vulnerabilities of autonomous vessels, smart buoys, satellite networks, and electronic customs systems have a high threat to supply chains globally. This highlights the importance of cybersecurity, regulatory control, and legal frameworks that have the ability to secure digital maritime infrastructure. Theoretically, smart ports and cyber-maritime systems show how technological innovation restructures logistics, risk management, and environmental management at the system level, portraying the relationship between the digital transformation and ethical custodianship.

Table 1. Conceptual Model of the Theoretical Framework

Pillar	Key Components	Main Outcomes
AI & Digital Oceans	AI, IoT, marine robotics, predictive analytics	Real-time monitoring, early threat detection, ecological resilience
Data-Driven Marine Systems	Blockchain, digital twins, interoperable data platforms	Transparency, traceability, integrated decision-making
Inclusive Stewardship	Women, coastal communities, indigenous groups	Social justice, community empowerment, participatory governance
Blue Finance	FinTech, blue bonds, carbon-credit platforms	Ethical markets, equitable investment flows, sustainable growth
Smart Maritime Systems	Autonomous ports, AI logistics, digital supply chains	Low-carbon operations, efficiency, secure cyber-infrastructure

3. METHODOLOGY

The proposed study will use a qualitative, interpretive research design to examine the potential of AI based marine intelligence, inclusive governance and community based stewardship as mutually enhancing drivers of sustainable change in the blue economy. The design that was selected is suitable since the main goal of the chapter is to draw conceptual understanding, meaning structures, and governance dynamics and not to measure causal relationships. It prefigures hermeneutical thinking, which allows the researcher to draw conclusions out of a convoluted of disciplinary cultures, including technology, governance, environmental science, and social equity, and reconstruct its applicability to sustainable ocean futures. This methodology corresponds to the modern demand of interdisciplinary maritime investigation, incorporating digital technologies in an ecological and social framework, and, consequently, being able to conduct a holistic analysis of sustainability (Ghosh and Kumar, 2025; Elston et al., 2024). The study has evidence base on a systematic literature review with thematic analysis and conceptual synthesis to supplement this review. To focus on peer-reviewed journal articles, scholarly monographs, and other authoritative institutional sources that cover the topics of artificial intelligence, digital ocean systems, maritime resource governance, blockchain and big-data utilisation, and an inclusive blue-economy development, sources were carefully chosen. The corpus includes the recent research on ocean digitalisation (Alam et al., 2024; Mohanbabu and Vettriselvan, 2025a; Mohanbabu and Vettriselvan, 2025b), inclusive governance and the community (Meena et al., 2025), and AI-facilitated ecological resilience (Gesami and Nunoo, 2020). Such thematic areas made sure that the review was in line with the three pillars of the chapter, which were digital transformation of ocean systems, inclusive governance mechanisms, and ecologically attuned stewardship models.

To ensure the methodological rigour, the literature review was conducted in three consecutive stages. To define the conceptual boundaries of the field, the literature was first broadly scanned to allow the identification of recurrent themes in the technological innovation, governance dynamics, ocean stewardship, and socio-ecological equity. Second, a set of patterns, points of intersection, and implicit assumptions in the works under consideration were identified through a manual coding technique based on the best practises of applying thematic analysis to digital maritime studies (Bakker, 2022; Eyo et al., 2025). Third, the coded content was synthesised to produce conceptual connexions among the three pillars of the theoretical framework, and thus demonstrate interconnections, as opposed to discrete themes, between AI systems, governance structures, and agency of society. The interaction between technological interventions and human-centred stewardship was taken into consideration specifically in the development of research themes based on the points of view expressed by Eluri et. al (2025) and Gee (2025). This orientation makes technological innovation a mediating variable to the socio-ecological systems rather than a deterministic variable. Empirical studies of the concept of maritime digitalisation have emerged to note that the achievement of the same is dependent on the availability of data, the integration of regulations, and the empowerment of communities (Khaskheli et al., 2025; Rengamani et al., 2019). Thus, the approach combines the experience of regenerative intelligence frameworks (Gee, 2025), smart ocean governance (Bashir, 2024), and data-driven ocean literacy (Larkin et al., 2022), which provides a consistent perspective on assessing how AI-powered systems can enhance sustainability in the framework of inclusive governance systems.

The interpretive framework was modified to include ethical considerations and data-integrity issues. The problems of algorithm prejudice, unequal access to data, digital sovereignty, and exposure of communities living along the coasts to technologically mediated governance were critically evaluated. The AI in marine monitoring, according to Zaeri (2025) and Alam et al. (2024), ethical governance should be integrated into the systems-thinking frameworks, which are responsive to the local socio-economic conditions. These interests shaped the interpretive approach of the chapter by having the conceptual synthesis informed by the sensitivity to structural inequalities and the asymmetries of governance. Other sources of literature on the topic of digital governance (Trice et al., 2021; Khaskheli et al., 2025) further conceptualised the methodological basis of evaluating how new technologies transform the workings of institutions and stakeholder power. The methodological design is adaptive in nature. Since AI-based ocean systems evolve, interpretive qualitative approaches can offer a process to reevaluate on an ongoing basis because technologies, regulatory landscapes, and socio-political conditions change. Such flexibility will ensure that the chapter is able to react to new evidence and developments as they take place. In turn, the ultimate methodological architecture is a triangulated one, i.e., a systematic review, interpretive thematic analysis, and conceptual synthesis across several disciplinary strands. This framework underpins an empirically sound and methodologically consistent analysis of the development of intelligent ocean systems and inclusive governance structures to co-evolve to increase climate resilience, social equity and environmental regeneration, hence providing a platform on which subsequent analytical sections can build on the significance of these conceptual pillars as reflected in new maritime transformations. One of the most significant changes in the modern maritime development has been the global transformations towards digital, climate-oriented, and socially inclusive ocean governance. This part is a critical discussion of the way artificial intelligence (AI), big-data analytics, blockchain systems and community-centred stewardship models redefine the blue economy. Following the contemporary literature in the interdisciplinary area (Alam et al., 2024; Eluri et al., 2025; Bakker, 2022; Meena et al., 2025; Bhola et al., 2025; Khan et al., 2025), it has been concluded that technological transformation is not sufficient. It is only through the incorporation of digital innovation into fair governance frameworks, open data forms and inclusion in the decision making process that it can result in ecological regeneration and socio-economic resilience.

4.1 Technological Transformation: From Reactive to Predictive Blue Governance

The idea that digitalisation is a choice that is optional has become unsustainable in the present maritime discourse. The blue economy currently faces the growing climate risks, declining fish stocks, widespread oceanic pollution, and growing geopolitical demands to decarbonise. Autonomous ships, satellite-enhanced ocean surveillance systems, machine-learning fisheries predictors, carbon management dashboard systems, and automated port infrastructure are some of the AI-powered predictive systems that are increasingly changing the character of maritime governance, that is, no longer being reactive by nature but anticipatory (Gesami & Nunoo, 2024; Nyangon, 2025).

Statistical data all over the world testify to this urgency:

- Oceans add up to about 3-6 trillion every year to the world economy (OECD ranges).
- Shipment of goods constitutes 90 percent of world trade (UNCTAD).
- IUU fishing brings a loss of 4.3-10million tonnes per annum (FAO).
- The ports emit approximately 3 % of the total GHG globally, which is caused by vessels call and cargo processing.

These statistics explain why the old style of maritime management, which is based on manual checkups, disjointed databases and slow reporting functions, will no longer work. Predictive analytics using AI facilitates the measurement of coral bleaching, prediction of hydrodynamic variations, optimisation of shipping routes to reduce emissions, and prediction of fish stocks variation through probabilities. Use of digital twins in a port environment strengthens the infrastructure resiliency by simulating climate-stress environments. In such a way, AI leads to the alteration of the maritime decision process, where the post-hoc correction will be replaced by a real-time adaptive governance- which is a necessary development as the strength of climate variability increases.

4.2 Transparency and Accountability through Digital Governance Systems

The fact that the fixing power of artificial intelligence can be much deeper than operational efficiency is beyond doubt. Bashir (2024) notes that digital oceans promote unmatched openness and distributed accountability. Supply-chain tracking that is verified by the blockchain improves the catch traceability, showing the vessel movements, cargo volumes, and distribution networks in the non-changeable ledgers (Khan et al., 2025; Suresh Babu et al., 2025). In this way, the regulatory authorities and local cooperatives are provided with the powers to control the activity of IUU, evaluate the ecological compliance, and verify the claims of sustainability. Similarly, the digital-emission audits and IoT-waste-tracking systems would help ports shift to the principles of a circular economy. AI analytics identify leaks of waste, track emissions at terminals, and provide automatic reporting, which forms the core of blue-finance innovations guided by blockchain traceability (Meccijenene, 2025; Bhola et al., 2025).

Table 2. Transparency Gains from AI & Blockchain in Maritime Governance

Governance Function	Traditional Model	AI/Blockchain Model	Empirical Impact
Vessel Monitoring	Manual reporting	Satellite-AI detection	40–60% increase in IUU detection
Catch Documentation	Paper-based	Immutable blockchain logs	Reduces misreporting by ~30%
Port Emission Tracking	Periodic estimates	Continuous AI-sensor analytics	Improves compliance accuracy
Waste Management	Limited visibility	IoT waste-tracking systems	Enhances circular performance

The table explains how the use of artificial intelligence and blockchain technologies has transformed the hitherto opaque maritime processes and made them transparent. In the traditional governance systems, such phenomena like misreporting, loss of data, and slow updates are adversely affecting conservation efforts. Verification mechanisms that are anchored on AI can be used to break these negative behaviours by removing weak linkages across the value chain. Leading to this transparency is no longer a technological object, but rather a substantive instrument of governance.

4.3 Inclusive Governance and Community-Centred Stewardship

One of the key findings out of this discussion is that technological capability by itself does not guarantee fair or sustainable results. Community-based governance is also invaluable towards creating legitimacy, compliance and facilitating ad hoc management. Participatory frameworks, thus, propose the idea of women, indigenous people, small-scale fishers, and youth networks which can play a vested role in the decision-making process, rather than being viewed as a passive source of data (Meena et al., 2025; Eyo et al., 2025). This opinion is supported by empirical evidence:

- MPA based on community leaders has a higher compliance rate of between 10 to 20 percent compared to the one managed by governmental authorities alone.
- Women make up most of the workforce in the fisheries processing sector but do not hold any more than 15 per cent. in leaderships (FAO).
- In some areas citizen-science networks report 25-40 percent of near-shore pollution.

The AI-assisted citizen science fills the gap between the local ecological know-how and machine intelligence. Internal cameras can also be used by the inhabitants of the coasts to record any illegal dredging, dumping of waste or the deaths of fish using the mobile applications, thus complementing the satellite derived information. This hybrid governance system complements the granularity as well as the legitimacy of the marine intelligence.

4.4 Workforce Transformation and Blue-Digital Skills

With the further digitisation of the maritime industry, workforce organisations are changing accordingly. The automated cranes, AI-enabled logistic systems, predictive maintenance, and automated navigation systems all change the job requirements. Mohanbabu & Vettriselvan (2025a, 2025b) underline that despite the increased productivity of these innovations, employees can be displaced unless the problem of digital-skills gap is resolved. Female and young populations in coastal areas are the most vulnerable to being marginalised by the new AI-driven jobs.

Table 3. Emerging Blue-Digital Skills for the AI-Enabled Maritime Economy

Skill Category	Examples	Rationale
Data & AI Skills	Maritime analytics, ML forecasting	Core for predictive governance
Cyber-Security	Secure vessel/port networks	Rising cyber risks in automation
Ecological Intelligence	Bioindicator monitoring	Links ecological and digital competencies
Technical Digital Skills	IoT, drones, autonomous systems	Required for modern port operations
Social-Governance Skills	Digital ethics, community engagement	Essential for inclusive governance

The skill matrix above shows that the emerging blue economy has now required a hybridised form of knowledge-base, which is formed by technical, ecological, ethical and community-oriented skills. Without a deliberate human-capital investment, digitalisation can effectively entrench the systemic process of marginalisation..

4.5 Comparing Traditional and AI-Enabled Maritime Systems

To synthesise the conceptual transformation, the following comparative table outlines the shift in governance philosophy:

Table 4. Traditional and AI-Enabled Maritime Systems

Dimension	Traditional Maritime Model	AI-Enabled Blue Economy Model
Governance	Centralised, expert-driven	Distributed, participatory, data-augmented
Resource Use	Linear & extractive	Regenerative & circular
Workforce	Manual, labour-intensive	Digital-technical & ecological
Technology Role	Operational support	Predictive intelligence & autonomy
Transparency	Limited monitoring	Blockchain-verified traceability
Community Role	Marginal	Central to stewardship

Artificial intelligence does not only threaten the running of processes but also questions the culture, which has guided the management structures. The traditional hierarchies have also been replaced by data-sharing ecosystems and the model of participative governance. Lacking a concerted effort to incorporate a voice of communities, the digital systems have a risk of reinforcing the existing power inequity rather than eradicating them.

4.6 Regulatory Gaps, Data Sovereignty, and the Blue-Digital Divide

There are major governance risks that cannot be resolved. Khaskheli et al. (2025) and Zaeri (2025) have also stated that autonomous vessels create legal ambiguities in terms of fault and liability. Surveillance systems built using AI are also an additional privacy risk, and untrained machine-learning architectures can also give rise to algorithmic bias, such as predictions of fishing-quota that proportionately prefer industrial fleets. The poor digital policies also threaten to disenfranchise the indigenous knowledge, as a result of which culturally based stewardship practises are undermined. The other significant risk is the development of blue-digital divide. At the same time as the hi-tech economies are using AI-powered port systems, developing seaside nations struggle with access to only basic marine information. Devoid of institutional support across the board, this digital ocean might only reinforce, but not reduce, the geopolitical hierarchies.

Table 5. Indicators of the Blue-Digital Divide

Indicator	High-Capacity States	Developing Coastal States
Satellite Ocean Data Access	High	Low or intermittent
AI-Driven Port Automation	Widely deployed	Limited to pilot stages
Ocean-Data Policies	Mature	Nascent

continued on following page

Table 5. Continued

Indicator	High-Capacity States	Developing Coastal States
Cybersecurity Readiness	Strong	Weak
Digital Literacy in Coastal Communities	Moderate-high	Low

This gap highlights significant economic disparity in economic power and technological readiness as well as human capital. Devoid of intervention via permeable structures and methodical capacity-building, the path of digitalisation can confer instead of mitigating marginalisation.

4.7 Economic Value of AI-Enabled Blue Systems

In spite of the challenges outlined above, AI-based maritime systems have significant economic potential. Smart agriculture enhances output and reduces the occurrence of the disease; predictive vessel analytics reduce pollution and congestion and fuel expenses (Gesami and Nunoo, 2024; Alam et al., 2024). Ports embracing paradigms of the circular economy generate new sectors of the industry - such as waste-to-energy, marine biomass refining, and green hydrogen bunkering - and new jobs accompany them (Kumar et al., 2024; Verma and Sharma, 2024). Performance of the regulated by blockchain also opens the door to sustainability-linked bonds and ancillary blue-finance bonds. The production of ecological intelligence and financial traceability forms the core of emergent regenerative economic models as it predicts the net-positive ecological impacts as opposed to minimisation of harm. The analytic project shows that the digital ocean systems can bring both fairly distributed prosperity and ecological regeneration only in case AI innovation is balanced with a strong sense of governance ethics, active civic engagement, and socio-ecological accountability.

The future blue economy is, therefore, not a technology-only outcome that has been shaped by intelligent institutional designs, participatory structures of governance, and sophisticated computational intelligence. Under these auspices, it will be only under digital oceans that sustainable development will be supported.

- Responsible AI
- Inclusive policies
- Equitable data governance
- Community stewardship
- Generative economic models.

Unless these conditions are met, the process of digitalisation will have the risk of reinforcing existing extractive paradigms. On the other hand, in a situation where such premises are fulfilled, the blue economy can develop into a resilient, transparent, as well as equitable maritime future.

5. FUTURE RESEARCH DIRECTIONS

The emerging intersection of marine intelligence, digital infrastructures and community-focused stewardship presents both new opportunities to study, but at the same time it reveals serious gaps, which need to be filled in order to build a sustainable and fair blue economy. Considering the systemic characteristics of this economy the connexion of ecology, technology, finance, governance, and culture, future studies should not stay in the fields of research. Rather, scholarship must embrace multifaceted sociotechnical, ecological, legal and economic approaches that can be used to understand the interdependence of ocean systems. One of the significant areas that require further work is the issue of AI transparency, explainability, and ethics in marine governance. Despite the significant potential of machine-learning models, their behaviour on the highly variable marine environment is not well known. The ecological interactions of coastal ecosystems are seen as non-linear, change in species distributions and climate processes through feedbacks making it uncertain how predictive models can be adaptive to uncertainty. The urgency is the way the AI decision systems, including automated quota setting, ecological forecasting, or surveillance algorithms, can be developed to avoid prioritising industrial fleets or contributing to the maintenance of the status quo. Future research ought to examine participatory AI validation models where the fishing cooperatives, indigenous populations and other local users of resources participate in the interpretation of training information and assessing the model outputs thus making sure that the algorithmic decisions are context-sensitive and socially acceptable. The other priority is the socio-economic impact of maritime automation, especially in port and logistics. Maritime digitalisation can bring drastic changes to the labour structure as the recent implementations of AI-based vessel routing, crane automation, container inspection, and customer processing have demonstrated (Mohanbabu and Vettriselvan, 2025a; 2025b). There is a need to quantify employment changes and define the new skills demanded, as well as to create the reskilling routes, which will avoid the increasing inequalities between the coastal people. Special consideration should be made to the measurement of digital-capability disparities among women, youth, and artisanal communities -groups that are already underprivileged in the labour market in maritime. Research projects incorporated into universities and technical institutes of maritime might be a pilot project to introduce curriculum changes to incorporate

AI ethics, ocean robotics, data literacy, and autonomous-systems maintenance into basic ocean-science training.

The issue of blue data sovereignty and digital ocean governance has to be addressed in future studies as well. Digital ocean commons will store sensitive datasets of vessel movements, seabed resources, biodiversity trends, and port- security systems and this will create problems of ownership, access, and equitable sharing of digital benefits. States and small island countries developing along the coasts are vulnerable to digital colonialism, in case data infrastructure is defined only by technologically dominant states or companies. The investigation is necessary concerning blockchain-based access procedures, federated data-sharing frameworks, and multilateral ocean-data treaties, which are in line with UNCLOS. Legal literature should also focus on the problem of algorithmic accountability, breaches of cybersecurity in autonomous maritime networks, and privacy limitations in satellite-based surveillance. Another important frontier is climate resilience. With increasing disruptions, including, but not limited to, rise in sea level, acidification, coral bleaching and extreme weather, AI-based real-time analytics will be the core of adaptive governance. Despite the fact that researchers like Nyangon (2025) and Gesami and Nunoo (2024) describe the emergent applications in marine-climate intelligence, more interdisciplinary efforts should be done to create multi-hazard AI-based coastal simulation models. These must include socio-economic vulnerability indices, ecological stress indicators and infrastructure fragility indicators. The co-design of such models and communities in the coastal areas would mean that experiential, gendered, and livelihood patterns become part of the digital resilience planning.

One new field of enquiry is the Regenerative Maritime Economics that not only focuses on sustainability but also aims at realising the ecological renewal as a source of economic value. Future studies must look into the ways AI can assist in regenerative marine agriculture- seaweed farming and shellfish bioremediation, community-owned networks of marine conservation and marine bioplastics circular factories. Although Gee (2025) offers a primary theoretical framework of the regenerative intelligence, empirical models of the association between technological adoption and ecological regeneration and social equity are still developing. Pilot studies might be carried out over long-term periods, i.e., observing how regenerative aquaculture performance, testing carbon -negative shipping fuels, or observing biodiversity restoration in digitally monitored MPAs, might help. The other agenda that is promising is the measurement of inclusive governance and ecological outcomes. There is an indication that gender inclusivity improves sustainability outcomes (Meena et al., 2025), but the strong comparative analysis studies are required to measure these associations. Mixed-method research, which is a strategic blend of both statistical modelling and ethnography, may be applied to find out the impact of gender-balanced leadership on compliance norms, rates of innovation, and patterns of resource-use in fisheries, port management, and start-ups in the maritime. Secondly, it is important to conduct studies to assess the fintech-enabled inclusion initiatives such as blockchain-validated microfinance solutions to empower women to participate in fisheries and ocean-technology initiatives (Khan et al., 2025; Suresh Babu et al., 2025).

Digital ocean systems gaps also include scalability and interoperability. Although numerous countries run AI-related maritime projects, not many of them have deployed them on national or regional levels. The barriers should be based on comparative analysis to determine the barriers based on infrastructure limitation, regulatory fragmentation, financing constraints, and cultural resistance. Studies might come up with national blue-AI preparedness measures of digital readiness, port-technology maturity, ocean-data strategies, human-capability, and gender mainstreaming. System-dynamics modelling could be also used to simulate the effect of incremental digital upgrades on emissions, employment, ecosystem health, and economic productivity. Lastly, blue-green digital convergence in terms of sectors should be studied in future. Offshore wind, marine hydrogen, tidal energy, and ocean-power systems are progressively connected to the grids of terrestrial renewable-energy sources, which require the integrated optimisation procedures based on AI. Circular port-energy clusters, blockchain-based carbon-credit exchanges, and hybrid ocean-energy systems are the ones that should be considered multidisciplinarily. A moral issue of seabed-mineral mining, deep-sea biodiversity conservation, and ocean-carbon-crediting which is scientifically proved requires immediate academic concern as well. Together, these research orientations highlight the necessity of interdisciplinary, ethically justifiable, and community-based scholarship that can help guide the shift towards a resilient, inclusive, and regenerative ocean governance.

6. CONCLUSION

The shift towards a digitally empowered, socially inclusive yet ecologically regenerative system, i.e., the blue economy, is one of the most impactful changes of the twenty-first century. As it is shown across the current chapter, the intersection between AI-based marine intelligence and community-based stewardship is not a non-essential addition to the maritime regulation but a necessary solution to the growing climate pressures, the loss of bio-diversity, and the structural inequalities in the ocean-based economies. These sustainable ocean futures cannot be created by the use of technology solely, they need the coordination of the digital capacity with the social inclusion, ethical governance and ecological reciprocity. Ocean should therefore not be seen as an economic frontier but as a living body of knowledge whose wellbeing is the foundation of climate stability, food security and wellbeing of human beings. It was noted that the analysis revealed the importance of artificial intelligence, big-data analytics, and blockchain in changing the way ocean resources are tracked, controlled, and replenished. Novel digital technologies, such as automated fisheries surveillance technologies, AI-assisted port management, and ocean-ecosystem digital twins, bring in unmatched conservation, prediction, and supply-chain visibility power (Alam et al., 2024; Nyangon, 2025; Gesami and Nunoo, 2024). These technologies can be used to hasten the ecological monitoring process, aid in responding to crises related to climate change, and improve the integrity of seafood and maritime logistics when properly used. However, one of the key lessons of this chapter is that technology innovation is no temporary replacement of fair governance. Devoid of institutional protection, citizen involvement, and ethical standards, even sophisticated digital systems are prone to widen the current inequalities, excluding traditional and indigenous knowledge, and enhancing control over ocean resources. Throughout the chapter, all the evidence supported the fact that social equity and gender inclusion were strategic forces of resilient maritime systems, as opposed to peripheral issues. Women and coastal communities are important sources of knowledge and economic power in fisheries, aquaculture, port logistics, and coastal tourism but they still experience institutional obstacles to finance, technology, education and leadership. The practical experiences of Meena et al. (2025) and Eyo et al. (2025) demonstrate that inclusive governance empowers compliance cultures, facilitates innovation, and generates more dynamic environmental results. Gender responsive budgeting in marine management, digital skills in marine management, and co-management in marine management arrangements that formally incorporate community leadership are therefore needed to institutionalise the marine stewardship. Inclusion is not just a moral requirement, it is the key to sustainability and good governance in the long term.

The strategic possibilities of circular-economy models in transforming the port systems and maritime logistics were also pointed out in the chapter. Circularity sees the creation of economic prosperity as a regenerative process, where the core capabilities are renewable energy, the valorisation of waste, lean shipping and digital traceability. Circular ports will be able to mitigate pollution, enhance competitiveness and create new markets in biofuels, biomaterials, and blue-tech innovation, as argued by Bhola et al. (2025) and Suresh Babu et al. (2025). Previously used as industrial logistics centers, ports are becoming more of a laboratory of climatic action and a center of ecological regeneration. To scale such innovations, however, will need investment in digital monitoring, inclusive policy instruments, and financing models that induce long-term ecological performance as opposed to short-term throughput. The other important contribution of the chapter is that it focuses on data governance and ocean-digital sovereignty as new spheres of strategic importance. Although digital oceans present a chance of transformations, it also comes with threats associated with privacy, cybersecurity, equity, and power concentration geographically. The data available about the ocean such as seabed structure, migration patterns, biodiversity patterns as well as climate patterns is of invaluable strategic value. To address the causes of digital ocean inequity, it is crucial to ensure agencies of coastal communities and the new maritime states can control their data. Since, future governance frameworks, as Zaeri (2025) and Khaskheli et al. (2025) suggest, need to establish, several specific accountability requirements, decentralised data-access, and safe and interoperative sharing protocols, must be defined. Moral digital governance is key to making sure that technological authority works in the best interest of the world and does not monopolise local knowledge systems but instead ensures the opposite.

Finally, the chapter proves that the future of the blue economy will depend not only on technological innovation but also on the level of leaders, the inclusivity of governance and the ecological foresight. In a future vision of ocean governance, there are AI systems that detect the vulnerability of coral-reef zones, blockchain tracked sustainable catches and climate-finance payments, women-led coastal businesses, AI-assisted marine zoning based on indigenous knowledge, and ports that are closed-loop energy systems. This direction goes beyond sustainability to regenerative maritime civilisation based on ocean ethics, technological humility and collective stewardship. To actualize this vision, it will be necessary that action be taken in harmony between technological, institutional, financial and even educational sectors. It requires collaboration between the public and the private, international cooperation in science and more robust institutional ability to manage more complicated maritime systems. More fundamentally it demands a cultural transformation: the re-branding of the ocean as not an extractive frontier but something like an ethical common-place of planetary health. With society engaging the ocean as a co-evolutionary companion instead of a depository reserve, a blue economy can be created that will not only be sustaining, it will also enrich the ecological and human prosperity. The meeting of AI, human values, and ocean wisdom is the start of a new era of maritime existence, where digital systems become the providers of regeneration and not supremacy. The success of this future would require unity, diversity, and a long-term dedication to the generations who will take over the oceans that we are shaping today.

REFERENCES

Alam, M., Khan, I. R., Siddiqui, F., & Alam, M. A. (2024). Artificial Intelligence as Key Enabler for Safeguarding the Marine Resources. In *Artificial Intelligence and Edge Computing for Sustainable Ocean Health* (pp. 409–451). Springer Nature Switzerland. DOI: 10.1007/978-3-031-64642-3_18

Bakker, K. (2022). Smart Oceans: Artificial intelligence and marine protected area governance. *Earth System Governance*, *13*, 100141. DOI: 10.1016/j.esg.2022.100141

Bashir, O. (2024). Harnessing Digitalization in Maritime Governance: A Systematic Review Strategic Challenges and Opportunities for Bangladesh. *Available at SSRN* 5125317.

Bhola, R., Aggarwal, S., Dhawan, P., & Mečėjienė, G. (2025). Blue Economy and Big Data Analytics. In *FinTech for Sustainable Finance and a Well-Blue Economy* (pp. 197–209). Springer Nature Switzerland. DOI: 10.1007/978-3-031-92390-6_10

Elston, J., Pinto, H., & Nogueira, C. (2024). Tides of change for a sustainable blue economy: A systematic literature review of innovation in maritime activities. *Sustainability (Basel)*, *16*(24), 11141. DOI: 10.3390/su162411141

Eluri, P. R. R., Annapurnaiah, K., Uday Bhaskar, T. V. S., Kiran Kumar, N., Padmanabham, J., Karthik, T. N. C., Vighneshwar, S. P., Reddem, V. S., Venugopala Rao, V., Murali Krishna, A., Ganesh, M. V., Lotliker, A., Nagaraja Kumar, M., Francis, P. A., Joseph, S., Balakrishnan Nair, T. M., & Srinivasa Kumar, T. (2025). Empowering the Blue Economy: The role of ICT in ocean observations, data, information and advisory services. *CSI Transactions on ICT*, *13*(1), 3–25. DOI: 10.1007/s40012-025-00407-x

Eyo, V. O., Eriegha, O. J., Andem, A. B., Joseph, A., & Ovioma, G. O. (2025). Harnessing the Blue Economy for Innovative Fisheries Management in Nigeria: Sustainable Strategies and Innovations. In *Handbook of Sustainable Blue Economy* (pp. 1–25). Springer Nature Switzerland. DOI: 10.1007/978-3-031-32671-4_88-1

Gee, R. O. W. (2025). Greening the blue ocean: Leading systemic transformation with regenerative intelligence. *Earth Environmental Science Research & Review*, *8*(1), 01-27.

Gesami, B. K., & Nunoo, J. (2024). Artificial intelligence in marine ecosystem management: Addressing climate threats to Kenya's blue economy. *Frontiers in Marine Science*, *11*, 1404104. DOI: 10.3389/fmars.2024.1404104

Ghosh, O., & Kumar, B. (2025). An Overview of Maritime Research, Its Significance, Challenges, and Opportunities: Navigating the Future. *Research Methods for Advancing the Maritime Industry*, 71-94.

Khan, S., Katoch, R., & Mahendru, D. (2025). Blockchain and FinTech Synergies: Driving Sustainability in the Blue Economy. In *FinTech for Sustainable Finance and a Well-Blue Economy* (pp. 171–195). Springer Nature Switzerland. DOI: 10.1007/978-3-031-92390-6_9

Khaskheli, M. B., Zhao, Y., & Lai, Z. (2025). Sustainable Maritime Governance of Digital Technologies for Marine Economic Development and for Managing Challenges in Shipping Risk: Legal Policy and Marine Environmental Management. *Sustainability (Basel)*, *17*(21), 9526. DOI: 10.3390/su17219526

Kumar, P., Jose, A. A., & Rajagopal, A. R. (2024). Sustainable Use of Marine Resources Using Ocean Technologies. In *Sustainable Development Goals* (pp. 249–274). CRC Press. DOI: 10.1201/9781003468257-14

Larkin, K. E., Marsan, A. A., Tonné, N., Van Isacker, N., Collart, T., Delaney, C., & Calewaert, J. B. (2022). Connecting marine data to society. In *Ocean science data* (pp. 283–317). Elsevier. DOI: 10.1016/B978-0-12-823427-3.00003-7

Meena, G., Vettriselvan, R., Rajesh, D., & Velmurugan, P. R. (2025). Diversity and Inclusion: Harnessing the Power of Inclusivity for Business Success. In *Security and Strategy Models for Key-Solving Institutional Frameworks* (pp. 203-234). IGI Global Scientific Publishing.

Mohanbabu, S., & Vettriselvan, R. (2025a). Focusing supply chain and container terminal challenges from the lens of Indian industries' perspectives with proposed data science interlinked artificial intelligence model. *International Journal of Procurement Management*, *24*(1), 92–114. DOI: 10.1504/IJPM.2025.148001

Mohanbabu, S., & Vettriselvan, R. (2025b). Will machine learning resolve the issues in container management in ports and logistics industries? *International Journal of Process Management and Benchmarking*, *20*(4), 559–575. DOI: 10.1504/IJPMB.2025.147399

Nyangon, J. (2025). Physics informed neural networks for maritime energy systems and blue economy innovations. *Machine Learning: Earth*, *1*(1), 011002.

Rengamani, J., James, F. A., Srinivasan, R., Poongavanam, S., & Vettriselvan, R. (2019). Impact on the usage of electronic data interchange (EDI) on the international shipping business in Chennai. *International Journal of Engineering and Advanced Technology*, *8*(2), 418–422.

Suresh Babu, C. V., Asmaa Begum, M. S., & Karafolas, S. (2025). Blockchain and AI for Advancing Sustainability in the Blue Economy: Innovating Marine Resource Management and Environmental Stewardship. In *FinTech for Sustainable Finance and a Well-Blue Economy* (pp. 127–150). Springer Nature Switzerland. DOI: 10.1007/978-3-031-92390-6_7

Trice, A., Robbins, C., Philip, N., & Rumsey, M. (2021). *Challenges and opportunities for ocean data to advance conservation and management*. Ocean Conservancy. DOI: 10.15868/socialsector.43891

Venice, J. A., Vettriselvan, R., Jain, S., Madusudanan, K., & Aarthy, C. C. J. (2025). Performance Evaluation and Metrics in Blockchain Powered AI/ML: Data Analytics for Cognitive Internet of Things (CIoT). In Transforming Education With AI-Powered Personalized Learning (pp. 143-178). IGI Global Scientific Publishing.

Verma, S., & Sharma, M. (2024). Role of Digitalization in a Sustainable Blue Economy. International Journal of Teaching. *Learning and Education*, *4*(3), 618013.

Zaeri, N. (2025). Securing Sustainability Through Artificial Intelligence and Blue Economy. In *Securing Sustainable Futures Through Blue and Green Economies* (pp. 37–82). IGI Global Scientific Publishing. DOI: 10.4018/979-8-3693-7893-9.ch003

Chapter 3
Blockchain in Conserving Marine Resources and Combating Illegal Fishing

Beyza Güdek
https://orcid.org/0000-0002-7432-9234
Karadeniz Technical University, Turkey

ABSTRACT

IUU fishing poses a serious threat to the world's marine ecosystems, with approximately 20% of global fish catches being caught illegally. This situation leads to a decrease in marine biodiversity, ecosystem degradation, and damage to local fishing economies. Blockchain technology stands out with its transparency, immutability, and decentralization features as a solution to this problem, where traditional control methods are inadequate. This study aims to develop a blockchain-based marine resources tracking system that provides end-to-end traceability of the fishing supply chain. The system records data such as location, time, boat, and fisher information of the fish caught in an immutable manner, allowing consumers to verify the origin of the seafood they purchase. The contribution of the system developed within the scope of the study to the protection of marine ecosystems, the prevention of illegal fishing, and the development of strategies for its adoption throughout the sector will be evaluated, and a concrete digital infrastructure model will be presented.

INTRODUCTION

Illegal, unreported, and unregulated fishing stands out as one of the most serious environmental problems threatening global marine ecosystems. Illegal fishing is becoming increasingly serious worldwide. Illegal fishing leads to negative conse-

DOI: 10.4018/979-8-3373-1777-9.ch003

quences such as a decrease in biodiversity, disruption of ecosystem balance, food security of coastal communities, and the seizure of economic livelihoods. Although traditional surveillance tools such as VMS and AIS offer significant contributions to the solution of this problem, they have serious limitations in terms of transparency, accessibility, and real-time intervention capacity.

Despite international agreements and technological advances, the lack of data integrity, independent auditing, and effective enforcement mechanisms in existing surveillance infrastructures is striking. In this context, blockchain technology emerges as a strong alternative solution to the traceability problem in the fishing sector with its structural features such as immutability, transparency, and decentralization. Blockchain ensures that transactions are verifiably recorded in the entire supply chain from catch to consumer, thus preventing illegally caught products from entering legal markets. However, the potential of blockchain-based systems for real-time auditing and rule enforcement in this area has not yet been sufficiently evaluated.

This study aims to close the existing gaps in traceability and regulatory compliance by designing a blockchain-based marine resources tracking system. The developed system can automatically audit legal regulations such as fish species, geographical location, catch dates and daily quotas through smart contracts and block transactions in case of violation and create an event record. Supported by the Python interface and Web3.py interaction protocols, this structure provides user-friendly access while also providing technological integration capability.

In this respect, the study does not limit auditing to passive data recording only, but also demonstrates the applicability of proactive digital auditing systems. At the same time, it contributes to the support of sustainable fisheries with digital transformation by offering a digital alternative to the limitations of VMS/AIS systems. The model presented by the research not only contributes to academic literature but also aims to provide a functional and expandable digital monitoring infrastructure for policymakers and implementing institutions.

The solution proposed in this study addresses data integrity, independent auditing, and effective enforcement gaps in existing infrastructures, including a feature set that enforces rules on-chain through smart contracts and event logging. Indeed, the expected impact is to transition from passive logging to proactive auditing, strengthening traceability and institutional accountability. This impact will be monitored through in-system metrics. This can be observed through the blocking rate of violating transactions, the consistency of the accompanying event logs, and the reduction in detection and response time during compliance audits. The successful implementation of the application will be achieved when it is implemented in conjunction with governance principles, such as a clear definition of institutional authority and data responsibilities, the establishment of authentication and role-based access, and the preservation of a balance between privacy and transparency.

BACKGROUND

Illegal, Unreported and Unregulated (IUU) fishing has become one of the most serious environmental and economic problems threatening global marine ecosystems. According to the Food and Agriculture Organization of the United Nations (FAO), one in five fish worldwide is caught illegally, which not only endangers the survival of marine species but also directly impacts the food security and livelihoods of coastal communities (Pratiwi et al., 2024). Scientific studies have shown that illegal hunting activities are particularly concentrated in areas with low control capacity, such as tropical regions, Southeast Asia, and the coasts of Africa, causing rapid depletion of fish stocks and disruption of ecosystem balance in these regions (Drammeh, 2000; Ndiaye, 2011).

The environmental impact of IUU fishing is multifaceted. Illegal fishing and overfishing, especially of highly valued species such as the Patagonian toothfish, are causing drastic depletion of their natural stocks and threatening the survival of predator species higher up the food chain (Collins, 2008). In addition, illegal fishing uses fishing practices that are utilized for the destruction of important habitat such as coral reefs and sea grasses, thus causing the collapse of target fish and even entire ecosystems (Barbosa-Filho et al., 2020).

Along with this environmental degradation, the social and economic consequences are also significantly negative. Fish is the main source of protein for people in deprived countries, and illegal fishing hurts local communities, jeopardizing food security (Lestari et al., 2020). Not only is this threatened by nutritional problems but also affects people's jobs. Small-scale fishers cannot compete with pirate fishing boats that illegally procure cheap and large amounts of fish. This leads to job loss, poverty, and social problems among coastal communities (Suherman et al., 2020). In addition, some illegal fishing is conducted by organized criminal groups and leads to serious human rights abuses like modern slavery and forced labor (Longo et al., 2021).

To prevent this situation, states use traditional control tools such as port state controls, regional fisheries management organizations, and satellite monitoring systems. However, these systems are often inadequate and cannot provide effective control, especially in cross-border waters and in areas where control opportunities are limited (Bethel et al., 2021; Busilacchi et al., 2018). International agreements, especially the Port State Measures Agreement (PSMA), although they seem effective in theory, have shown limited success in many countries due to insufficient control capacity and political will in practice (Hagan, 2014).

This is where blockchain technology stands out as a digital solution. The immutable, transparent, and decentralized structure of blockchain allows every stage in the fishing supply chain to be recorded, from the moment of catch to the consumer.

This prevents illegally caught fish from mixing with legal markets and enables consumers to directly verify the source of the product (Pratiwi et al., 2024). Blockchain infrastructures that work integrated with carding systems and certification applications implemented in regions such as the European Union can also stop the flow of illegal products in foreign trade by providing automatic control at customs (D. E. Kim & Lim, 2024).

In light of all these assessments, it is clear that IUU fishing is not only an environmental threat but also a multifaceted problem that creates crises in many dimensions, such as human rights, economic justice, global food security, and maritime security. In solving this problem, the transparency and traceability offered by blockchain-based digital systems are a strong and sustainable alternative to close structural gaps that cannot be resolved with traditional methods. However, in order for these technologies to be successful, it is critical that they are supported by international policies, legal infrastructures are strengthened, and local stakeholders are included in the process. Only with this holistic approach can it be possible to permanently eliminate the damage that illegal fishing causes to nature, people, and the economy.

COMBATING ILLEGAL, UNREPORTED, AND UNREGULATED FISHING

Illegal, Unreported, And Unregulated Fishing as a Global Threat

IUU fishing has been identified by FAO as a serious threat to the sustainability of global marine resources since the 1990s (Food and agriculture organization, 2001). The devastating effects of these activities on marine ecosystems and food security have necessitated comprehensive measures at the international level (Lee et al., 2014). In this context, the International Plan of Action to Combat Illegal, Unreported and Unregulated Fishing (IPOA-IUU), adopted by FAO in 2001, provided the first holistic policy framework at the international level to combat this threat (Food and agriculture organization, 2001). IPOA-IUU has detailed the responsibilities of states as flag states, port states, and coastal states, and has emphasized the importance of measures such as technological monitoring systems and data sharing (Sodik, 2008). This document has also served as an important roadmap to combat IUU fishing through monitoring, surveillance, and control (MCS) mechanisms implemented by regional fisheries management organisations (RFMOs) (Palma-Robles, 2009).

Evolution of Monitoring and Tracking Systems: VMS and AIS

Vessel Monitoring System (VMS) and Automatic Identification System (AIS), two basic systems developed for the sustainable management of fishing activities and effective combat against IUU fishing, form the basis of surveillance and monitoring technologies in the maritime field.

VMS was initiated in the 1990s, managed by the European Union, and allows the real-time reporting of the satellite-determined position, speed, and direction of fishing vessels to monitoring authorities. The system utilizes satellite communication systems like Inmarsat, Argos, or Iridium for automatic reporting of data at regular intervals. With this, the movement patterns of the vessel can be examined, and fishing activity and behavior within allowed zones can be tracked. In the early 2000s, the use of VMS devices was made compulsory on all fishing vessels 15 meters and more in length within the European Union, and the system also enhanced efficiency in monitoring as well as resource management (Chuaysi & Kiattisin, 2020a; Longépé et al., 2018). In the USA, NOAA utilizes VMS systems in the Pacific and Atlantic fleets and combines them with quota and area-based management solutions, particularly for specific species (Gaspar et al., 2021; Y. H. Kim et al., 2014).

AIS is a system initially developed by IMO for maritime safety and collision prevention purposes and made mandatory for all commercial ships over 300 GT with the SOLAS (Safety of Life at Sea) agreement in 2002. Operating over VHF radio frequencies, AIS continuously broadcasts information such as the ship's identity (IMO number), location, speed, and route. The most important feature of AIS is that it is an open-access system. In this way, not only state authorities but also non-governmental organizations, academic research teams, and global data platforms (e.g., Global Fishing Watch, Marine Traffic) can analyze AIS data. This transparency plays an important role in the disclosure of IUU fishing and the protection of resources (Tetreault, 2005). In addition, by processing AIS data with big data technologies, ship behavior patterns can be extracted, and illegal fishing trends can be detected (Han et al., 2025).

Both systems possess pros and cons. VMS offers more secure and timely information, which makes it easier for the authorities to handle things more effectively. This is more convenient for the imposition of inspections and penalties. Its information is not publicly accessible. AIS is a better system for research and cooperation on an international level since it provides information more frequently and is open in nature. However, AIS openness also presents the possibility of abuse (e.g., false signals) (Shepperson et al., 2018). For this reason, the majority of countries are incorporating VMS and AIS systems to run alongside one another, constituting a multi-layered system for tracking as well as controlling maritime operations (D. E. Kim & Lim, 2024).

Technologies for monitoring ships, including the VMS and AIS, have gained global widespread application; however, there are great variations in application based on the inspection capacity of nations, their technological infrastructure, and political will. The European Union (EU), a pioneer region in this arena, mandated the fitting of VMS on fishing vessels longer than 15 meters from 2000 to 2005 (Longépé et al., 2018). The United States of America (USA) has been installing VMS and AIS in various areas with varying uses since 2003, introducing these technologies to coastal and offshore fleets, notably under NOAA's leadership (Tetreault, 2005). In comparison, Canada implemented VMS primarily for specific kinds of fisheries in the 2000s, which involved more restricted use of the system.

Indonesia has developed a model policy on transparency, which mandates the use of VMS by big fishing vessels starting from the year 2018. Apart from this, the data is made available publicly (Longépé et al., 2018). Norway has been using an overarching approach to monitor all merchant ships by combining AIS and radar data since 2005. The system is very successful in terms of maritime security and in the fight against IUU fishing (Shepperson et al., 2018). In Türkiye, with the Fisheries Regulation, VMS has been made compulsory for fishing vessels longer than 12 meters, as of 2009. This application is a significant move in Turkey's accession to regional fisheries management arrangements. Such discrepancies mirror not just the technological capacities of the nations involved but also the value they place on sustainable fishing, their policies of transparency, and the stringency of their regulating mechanisms (Gaspar et al., 2021).

While technological solutions such as Vessel Monitoring Systems (VMS) and Automatic Identification Systems (AIS) are useful instruments to address Illegal, Unreported, and Unregulated (IUU) fishing, such solutions in themselves are inadequate to fully eliminate the practice. Chief among the reasons is that the coverage of VMS systems is limited to authorized entities only, thereby excluding civil society organizations and independent monitoring, consequently constraining transparency and public accountability (Chuaysi & Kiattisin, 2020b). Whereas AIS systems provide a more open architecture, the potential for easy deactivation of equipment or intentional interference with signals "dark vessels" severely reduces the utility of this system (Welch et al., 2022). Furthermore, latency in transmission makes it difficult to implement real-time interventions, and thus undermines deterrence (Longépé et al., 2018).

Weaknesses at the legal and political level also constitute another important problem. While the sharing of authority in international waters is not clear, flag state practices are abused. In particular, the preference for flags of countries with low controls (flag of convenience) causes IUU activities to become widespread (Doherty et al., 2021). International agreements such as the PSMA are not sufficiently implemented by many developing countries (Phuong & Pomeroy, 2023).

Data integrity and traceability issues in the supply chain also create weaknesses in the fight against IUU fishing. Manipulation of documents in ports, uncertainty of cargo exchange points, and the inability to link VMS data to invoices or transaction documents prevent transparent chain control (Brown et al., 2024).

Although existing surveillance and tracking systems used in the literature to combat IUU fishing have made significant contributions, some fundamental deficiencies limit the effectiveness of these systems. First of all, the independent management of AIS, VMS and port data causes discontinuities and lack of data integration in the fishing supply chain (Chuaysi & Kiattisin, 2020b). Additionally, VMS data is largely kept private, even between states, which is a major problem that hinders international cooperation and transparency (Doherty et al., 2021). Although it is possible to detect illegal activities with existing technologies, the fact that these detections cannot be translated into rapid and effective legal or physical interventions represents another weakness in the implementation phase (Longépé et al., 2018). Another deficiency that is increasingly emphasized in the literature is that blockchain-based systems are not sufficiently applied in product tracking and certification processes (Abou Jaoude & George Saade, 2019; D. E. Kim & Lim, 2024). Finally, although AI-supported detection methods are theoretically promising, especially in the fields of image processing and anomaly analysis, they have not yet been integrated into large-scale systems (Brown et al., 2024).

This study proposes a solution to the gaps mentioned above, develops a blockchain-based distributed architecture, and proposes a reliable recording infrastructure that is impervious to manipulation, unlike centralized data structures. The system automatically detects rule violations through smart contracts and instantly notifies relevant stakeholders. This feature shows that not only passive monitoring, but also proactive intervention mechanisms can be implemented. Simulation results have demonstrated the success of the system in terms of data integrity, transparency, and rule application consistency, and it has also confirmed its potential to transform into a multi-layered marine monitoring infrastructure in the future with analysis modules and VMS/AIS integrations.

METHODOLOGY

This study aims to develop a traceability-based blockchain system to prevent IUU fishing and to test this system in a simulation environment. The developed system works with real-time data recording, rule-based control mechanisms, and a smart contract infrastructure that can generate automatic warnings (alerts) in case of violations. In this context, the components of the system, the tools used, the testing process, and verification mechanisms are explained in detail below.

System Components and Architecture

The developed system consists of four main components: (1) a Python interface that receives data from the user, (2) smart contracts that manage the catch records and audit rules, (3) an Ethereum-based blockchain environment running on the Hardhat local network, and (4) chain data where the system outputs are recorded. An architecture that visualizes the overall operation of the system is presented in Figure 1.

Figure 1. Blockchain-based system architecture

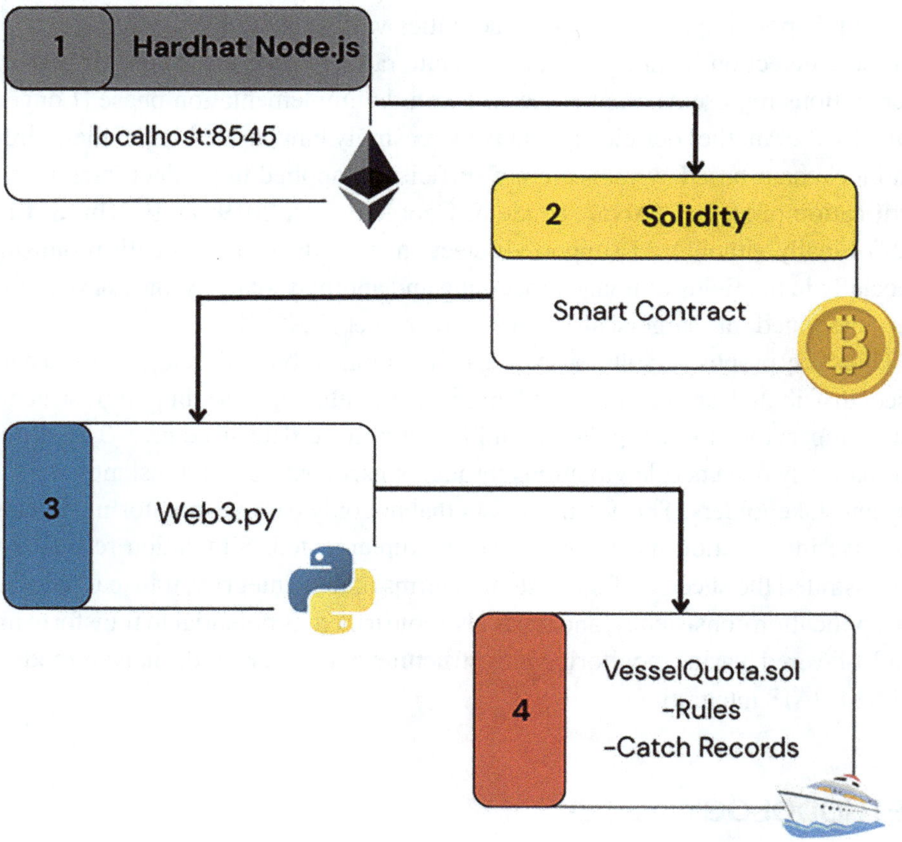

In this architecture, the hunting data received from the user is sent to smart contracts via Python, and the contracts both record this data and compare it with the rules to check for possible violations. All transactions are stored on the blockchain in an unchangeable manner and archived with a timestamp.

Development Environment and Technological Infrastructure

Open-source, blockchain-compatible tools were preferred during the development process. Hardhat was used for writing and testing smart contracts, and the Python programming language and Web3.py library was used on the interface side. The system is structured with the components in Table 1.

Table 1. Tools used for system development

Component	Description / Version
Node.js	JavaScript runtime
Hardhat	Smart contract framework (v2.20.0)
Solidity	Contract language (v0.8.28)
Python	Interface language (v3.10+)
Web3.py	Ethereum interaction (v6.x)

The local development environment was created with the npm init command, then Hardhat installation was performed with the commands in Code 1. The compilation of smart contracts was performed with the npx hardhat compile command.

Coding and Simulation

This section describes the step-by-step coding and simulation process of the proposed blockchain-based traceability system. Smart contracts were deployed and tested on the Hardhat native Ethereum network. Through the Python interface, catch data was transmitted to the smart contracts and queried from the chain. How the system works, how rule checking and data immutability are ensured, are demonstrated practically with code blocks.

Table 2. Code 1- Hardhat Installation

Code	*npm install --save-dev hardhat*
Output	*Compiled 3 Solidity files successfully*

One of the developed contracts, FishTrace.sol, was deployed to the local Ethereum network (localhost:8545) with the command in Code 2.

Table 3. Code 2- Native Ethereum Network Deployment

Code	*npx hardhat run scripts/deploy-fishtrace.js --network localhost*
Output	*Deployed FishTrace to: 0xDc64a140Aa3E981100a9becA4E685f962f0cF6C9*

This address was used by assigning it to the contract_address variable in the main2.py script on the Python side. The Hardhat local network provides 20 default test accounts. Since these accounts are "unlocked" by default, they allow direct transactions to be sent without signing. Thanks to this structure, the system's testing operations were carried out quickly and securely.

Smart Contract Design and Interaction with Python

The first smart contract developed, FishTrace.sol, is designed to record and track basic fishing data on the blockchain. Through this contract, information such as fish species, location information, the fisherman's name, and the vessel ID for each catch record is stored on the chain in an immutable manner. These records allow for transparent and auditable product tracking throughout the supply chain.

The data recording process in Code 3 is performed through the registerCatch() function defined in the contract. The function takes four basic parameters: fishType, location, fisherName, and vesselID. This information is automatically timestamped with block.timestamp and stored on the chain.

Table 4. Code 3- Function that provides data logging in FishTrace.sol contract

```
function registerCatch(
 string memory fishType,
 string memory location,
 string memory fisherName,
 string memory vesselID
) public {
fishRecords.push(FishRecord({
fishType: fishType,
location: location,
timestamp: block.timestamp,
fisherName: fisherName,
vesselID: vesselID,
processor: "",
distributor: "",
market: "",
consumer: "",
completed: false
}));
}
```

To provide external system interaction with this contract, the Python language and the Web3.py library were used. In the main2.py file, which is a Python script, the data received from the user was directly transmitted to the contract, and the registration process was initiated. Code 4 includes the registration process using sample data. As a result of this process, a record was added to the chain in line with the data sent to the contract, and the process was completed successfully.

Table 5. Code 4- Python code that performs data recording in main2.py

<table>
<tr><td>Code</td><td>

```
tx = contract.functions.registerCatch(
 "Mackerel",
 "Black Sea Coast",
 "John Fisher",
 "VESSEL-001"
).build_transaction({
 "from": account,
 "gas": 2000000,
 "gasPrice": w3.to_wei("1", "gwei"),
 "nonce": w3.eth.get_transaction_count(account)
})
tx_hash = w3.eth.send_transaction(tx)
receipt = w3.eth.wait_for_transaction_receipt(tx_hash)
print(f" Fish catch recorded. Transaction: {tx_hash.hex()}")
```

</td></tr>
<tr><td>Output</td><td>Fish catch recorded. Transaction: 0x1d402abea31f3ac6801e9e39b5fa...</td></tr>
</table>

To verify the record, the record on the blockchain can be queried using the getFishRecord() function defined in the contract. This function in Code 5 returns the data corresponding to the record number to the user.

Table 6. Code 5- Reading registered fish data from blockchain

Code	data = contract.functions.getFishRecord(0).call() print("\n Fish Trace Record:") print(f'Type: {data[0]}") print(f'Location: {data[1]}") print(f'Timestamp: {data[2]}") print(f'Fisher: {data[3]}") print(f'Vessel ID: {data[4]}") print(f'Processor: {data[5]}") print(f'Distributor: {data[6]}") print(f'Market: {data[7]}") print(f'Consumer: {data[8]}") print(f'Completed: {data[9]}")
Output	*Fish Trace Record:* *Type: Mackerel* *Location: Black Sea Coast* *Timestamp: 1716409183* *Fisher: John Fisher* *Vessel ID: VESSEL-001* *Processor:* *Distributor:* *Market:* *Consumer:* *Completed: False*

When the above code is run, all information about the relevant record is presented to the user on the terminal screen. Thanks to this structure, the system not only records data; it also allows this data to be stored and tracked in an auditable manner. The record structure in question provides a tracking function in accordance with the principles of reliability and transparency at every stage of the supply chain.

Rule Based Control Mechanism

The second main smart contract of the system, VesselQuota.sol, was developed to ensure that fishing vessels fish within the specified legal limits. This contract allows for the definition of rules specific to each vessel and the automatic auditing of these rules by comparing them with the catch records. When a transaction against the rules is detected, the system cancels the transaction (reverts) and records the violation by issuing an event called Alert on the chain.

Defining a Rule

The rules that will be valid for each boat are defined via the addRule() function. These rules, given in Code 6, include variables such as valid date range (startDate, endDate), allowed fish type (fishType), geographical limits (minLat, maxLat, minLong, maxLong), and daily maximum catch amount (dailyLimit).

Table 7. Code 6- The addRule() function which defines boat-specific rules

```
function addRule(
  string memory vesselID,
  string memory fishType,
  uint256 startDate,
  uint256 endDate,
  int256 minLat,
  int256 maxLat,
  int256 minLong,
  int256 maxLong,
  uint256 dailyLimit
) public {
  vesselRules[vesselID] = Rule({
  fishType: fishType,
  startDate: startDate,
  endDate: endDate,
  minLat: minLat,
  maxLat: maxLat,
  minLong: minLong,
  maxLong: maxLong,
  dailyLimit: dailyLimit });}
```

Thanks to this structure, rules can be defined for different geographical areas and different types of fish for each boat. Once the rules are recorded on the blockchain, they cannot be changed from the outside.

Rule Control and Data Recording

Catch records are performed with the recordCatch() function. Before recording data to the chain, this function compares the entered data with the rules previously determined with the addRule() function. The function structure that performs this operation is included in Code 7.

Table 8. Code 7- Rule control and suspension of processing in case of violation

```
function recordCatch(
 string memory vesselID,
 string memory fishType,
 int256 latitude,
 int256 longitude,
 uint256 weight
) public {
 Rule memory rule = vesselRules[vesselID];
 require(block.timestamp >= rule.startDate && block.timestamp <= rule.endDate, "Invalid date");
 require(keccak256(bytes(fishType)) == keccak256(bytes(rule.fishType)), "Invalid fish type");
 require(latitude >= rule.minLat && latitude <= rule.maxLat, "Invalid latitude");
 require(longitude >= rule.minLong && longitude <= rule.maxLong, "Invalid longitude");
 require(dailyCatch[vesselID][getToday()] + weight <= rule.dailyLimit, "Daily quota exceeded");

 dailyCatch[vesselID][getToday()] += weight;
 emit CatchRecorded(vesselID, fishType, latitude, longitude, weight, block.timestamp);}
```

If any of the above rules are violated, the transaction is stopped (reverted) with the require command before it is written to the chain, and the violation is detected. In this way, an automatic audit is performed at the contract level.

Violation Scenario

In order to verify the ability of the developed smart contract to perform compliance audits, a special test script called test-violation.js was created and various scenarios were systematically simulated through this script. This testing process is critical to assess whether the contract correctly recognizes violation scenarios.

The scenarios covered in the simulations include the following situations:

- *Performing fishing activities outside the allowed time intervals:* It is expected that every fishing operation performed outside the fishing periods specified in the smart contract will be rejected by the system. In this way, seasonal sustainability principles are preserved.
- *Transactions with a fish other than the defined species:* Transactions can only be made with predefined and allowed fish species in the system. In case of any data entry other than these species, the contract's verification mechanism is activated and cancels the transaction.
- *Fishing outside the defined geographical boundaries (e.g., lat-long coordinates):* Geographic boundaries are determined by locational parameters fixed in the contract. Data of vessels fishing outside these boundaries are rejected without being recorded on the chain. One of the sample scenarios where this situation is tested is covered in the test called Code 8. In the relevant test, an attempt was made to record fishing with a coordinate outside the defined

area, and as expected, the system considered this transaction invalid. When the test is run, the transaction in question is not recorded in the chain; an error message automatically generated by the contract is displayed in the terminal.
- *Exceeding the daily catch quota:* The maximum amount that can be caught per day for each boat is limited. In case of a transaction exceeding this limit, the contract considers this situation as a violation and does not record the relevant transaction.

These scenarios clearly show that the system only approves and records valid and legal fishing activities in the chain; violations are detected and rejected. Thus, the effectiveness of the system in terms of both data security and compliance with sustainability principles has been tested.

Table 9. Code 8- Violation scenario in the test-violation.js script

Code	await vesselQuota.addRule("VESSEL-001", "Hamsi", start, end, 40, 45, 30, 35, 100); await expect(vesselQuota.recordCatch("VESSEL-001", "Hamsi", 50, 50, 20)).to.be.revertedWith("Invalid latitude");
Output	*Error: VM Exception while processing transaction: revert Invalid latitude*

This mechanism ensures that all transactions that violate the rules are stopped at the chain level and violations are documented with log records. In addition, the contract includes an event definition called Alert, which can be triggered at the time of the violation and tracked by external systems.

Simulation and System Outputs

The functionality of the developed system and the correctness of the rule-checking mechanisms were tested on a local simulation environment. In the simulation process, the Ethereum test network (localhost:8545) provided by Hardhat was used, and all transactions were performed on this network. This network provides 20 default test accounts, and since each of them has unlimited test Ether, transactions can be carried out without requiring signing.

Daily Catch Record and Total Catch Quantity Inquiry

The main.py script developed on the Python side transmits the amount of fish received from the user to the SmartFishing.sol contract and queries the total daily catch. Code 9 ensures that a 25 kg catch record is transmitted to the contract. When this code block is run, first the record is recorded on the chain via the recordCatch() function, then the daily total is queried with the getCatch() function.

Table 10. Code 9- Hunting record and total query process in main.py

<table>
<tr><td>Code</td><td>

```
def record_catch(amount_kg):
nonce = w3.eth.get_transaction_count(account)
tx = contract.functions.recordCatch(amount_kg).build_transaction({
'from': account,
'gas': 2000000,
'gasPrice': w3.to_wei('1', 'gwei'),
'nonce': nonce
})
tx_hash = w3.eth.send_transaction(tx)
receipt = w3.eth.wait_for_transaction_receipt(tx_hash)
print(f" {amount_kg} kg of fish recorded. Tx hash: {tx_hash.hex()}")

def get_today_catch():
today = int(time.time() // (24 * 60 * 60)) # UTC day number
catch = contract.functions.getCatch(account, today).call()
print(f" Total catch for today: {catch} kg")
```

</td></tr>
<tr><td>Output</td><td>25 kg of fish recorded. Tx hash: 0xd87e4dff28fa1c78...
Total catch for today: 25 kg</td></tr>
</table>

This result shows that the system successfully performs both registration and query operations. The user can verify only their total catch amount each day from the blockchain with the getCatch() function. This mechanism makes it easier to detect limit exceedances of the registrations made on behalf of the same boat or user.

Recording and Event Tracking with Timestamp

All records are automatically timestamped with the block.timestamp parameter. Thanks to this mechanism, each catch record is recorded in the chain with not only the quantity and location, but also the time it occurred. Code 10 shows how this timestamp is used in FishTrace.sol.

Table 11. Code 10- Adding the catch record to the blockchain with a timestamp

Code	timestamp: block.timestamp
Output	*Timestamp: 1716409183*

This record is then delivered to the user when queried via the getFishRecord() function. This process demonstrates that the system maintains record integrity and is transparent to external audits. This value can be converted to date and time format in external systems such as data visualization dashboards. This allows time-based analysis of records.

Events Generated in the Chain

In case of a violation, an event called Alert is triggered in the VesselQuota.sol contract. These events can be captured by blockchain listeners and integrated into instant alert systems. For example, an event definition like in Code 11 is used.

Table 12. Code 11- Event declaration in VesselQuota.sol

```
event Alert(string vesselID, string reason, uint256 timestamp);
```

Thanks to this structure, violations can be transmitted not only as transaction reversals but also outside the system in log format. This transforms the system into a decentralized control mechanism.

Key Management, Security and Restrictions

The security and integrity of the developed blockchain-based system are provided through both the immutable structure of Ethereum and the rules of smart contracts. In this section, the key management policy implemented in the test environment, security advantages, and current limitations of the system are discussed.

Key Management and Transaction Authorization

The Hardhat local network used in the simulation environment provides 20 default test accounts. These accounts work in an unlocked form to send transactions on the chain and do not require a separate private key management. This structure offers developers an easy and fast testing opportunity without a transaction signing process.

In Code 12, the Python command used as an example of this situation shows direct transaction sending without using any private key.

Table 13. Code 12- Sending unsigned transactions in a test environment

```
tx_hash = w3.eth.send_transaction(tx)
```

This command allows transactions to be made on the test network by specifying only the from address. The from statement identifies the account that sent the transaction. Normally, to send a transaction in blockchain systems such as Ethereum, the transaction must be digitally signed with the private key of this account. However, in local test networks such as Hardhat, since test accounts are provided unlocked by default, transactions can be made directly without having to access the private keys of these addresses. In other words, the transaction can be sent to the network by specifying only the from address with the send_transaction(tx) command. This is not possible in real networks for security reasons and is a convenience that is only valid in local simulation environments.

Implementation Differences in Real Networks

In case the developed system is integrated into the real Ethereum network or special blockchain infrastructures by removing it from the local simulation environment, various technical arrangements will be needed. First of all, to securely sign every transaction to be performed in real networks, private keys belonging to the user must be used, and it is of great importance that these keys are stored securely. In addition, service providers such as Infura, Alchemy, or specially configured RPC endpoints can be used to connect to a decentralized node network instead of a local network. To provide such a connection, an HTTPS-based node connection must be defined via the Web3 client as given in Code 13, and access must be provided through this structure. While these changes enable the system to work on a decentralized network, they will also allow a more advanced structure to be provided in terms of security, performance, and scalability.

Table 14. Code 13- Node provider example for real network

```
w3 = Web3(Web3.HTTPProvider("https://mainnet.infura.io/v3/YOUR_PROJECT_ID"))
```

In real networks, every transaction sent incurs a gas cost and cannot be reversed once recorded on the chain. Therefore, additional measures are required for record accuracy, secure authentication, and data integrity.

RESULTS

In this study, the blockchain-based traceability system developed to prevent IUU fishing has demonstrated its potential to provide transparency, reliability, and digital audit capacity in the seafood supply chain. The system undertakes the basic data recording function with the smart contract called FishTrace.sol, where information such as fish species, location, catch date, fisherman name, and vessel ID is recorded on the chain. These records are stored in an unchangeable manner due to the structure of the blockchain, and each transaction is archived with a timestamp. This feature prevents data manipulation, ensures the accuracy of the records, and provides verifiable evidence for external audit bodies.

The second component of the system, the VesselQuota.sol contract, allows the definition of geographical, temporal, and species-based fishing rules specific to each vessel and the automatic auditing of these rules. During the simulation process, fishing attempts against the determined rules were canceled (reverted) by the system without being recorded on the chain, and these violations were recorded as events on the blockchain. Thus, the system functioned not only as a passive recording infrastructure but also as an active auditing and breach detection mechanism.

The user interface developed in Python successfully performed both data recording and querying operations by interacting with smart contracts via Web3.py. Each catch record entered by the user was sent to the chain, and then performance tests were conducted by querying the total daily catch amount. The outputs obtained in the tests revealed that the system has a high capacity for simultaneous transaction execution and verification.

In addition, the fact that the transaction history can be traced retrospectively with the timestamps recorded in the chain made it possible for both regulatory institutions and consumers to establish trust at every step of the supply chain. The Alert event mechanism defined in the system provides a warning infrastructure that can be used for integration with external systems in the event of a potential breach. This feature forms the basis of decentralized digital audit systems.

However, it should be noted that the system currently only works on a local test network (localhost:8545), does not have a graphical user interface (GUI), and does not integrate the private key management and RPC connections required to work on real networks. These limitations limit the applicability of the system on an industrial scale; however, it is anticipated that these limitations can be overcome as the development and integration process continues.

This developed system effectively uses the immutability, transparency, and decentralization advantages offered by blockchain technology, providing traceability in the seafood supply chain and providing a viable digital solution in the fight against IUU fishing.

CONCLUSION AND DISCUSSION

This study presents alternative solutions that digital technologies can provide against the shortcomings of traditional methods used in combating IUU fishing through a comprehensive model. Traditional monitoring and tracking systems, such as VMS and AIS, which are widely included in the literature, provide significant contributions in terms of location and fleet tracking, but face some serious limitations in practice. In particular, the fact that AIS signals can be easily disabled and VMS data can only be accessed by authorized authorities causes these systems to be weak in terms of public control and transparency (Chuaysi & Kiattisin, 2020b; Welch et al., 2022).

The developed blockchain-based system is designed to compensate for the shortcomings of these traditional methods. Blockchain technology, which is decentralized and provides equal access among stakeholders, significantly reduces the limitations of existing systems by making the entire process transparent and verifiable. As stated by Kim & Lim (2024) blockchain-based traceability systems not only keep records of fishing activities but also play an important role in preventing counterfeiting and ensuring food safety by covering all stages from processing to distribution and delivery of products to the end consumer.

One of the most critical features of the blockchain infrastructure used in this study is that the data is immutable. Since it is not possible to delete or change the data once entered into the chain, this significantly increases the security and integrity of the records (Pratiwi et al., 2024). Each transaction performed in the system is automatically sealed with a timestamp (block. timestamp), and the chronological accuracy of the transactions is ensured. Thus, all data in the chain can be openly audited by third parties, and the principle of transparency is effectively supported throughout the system.

Thanks to the smart contracts developed in the study, hunting activities are subject to automatic auditing by comparing them with the determined rules. As emphasized in the studies of (Longépé et al., 2018) and Brown et al. (2024), it is not only sufficient to detect illegal activities, but also to be able to stop these activities immediately is of critical importance. In this direction, the system developed automatically detects and stops rule violations through smart contracts and brings the concept of "proactive digital audit systems", which have limited application in the literature, into practice.

In addition, a large part of the existing systems stores fishing data centrally, which makes the data vulnerable to manipulation (Chuaysi & Kiattisin, 2020b). The blockchain infrastructure developed within the scope of this study ensures that data is stored securely and transparently, thanks to its decentralized, distributed structure. In addition, the system proactively informs controllers and stakeholders with an event-based notification mechanism in case of possible violations.

Despite all these advantages of the system, some technical and operational limitations should also be taken into account. Currently, the system has only been tested in a local simulation environment and has not yet been integrated into private blockchain infrastructures via the real Ethereum network or RPC providers such as Infura. In this context, the performance, cost, and scalability issues that the system will encounter in real-world applications still require evaluation. In addition, the lack of a graphical user interface (GUI) and the fact that IPFS-based long-term data storage solutions have not yet been integrated to limit the usability of the system for end users.

In the future, with the elimination of these limitations, the integration of the system with real-time VMS/AIS data and AI-supported analysis modules will play a critical role in the creation of comprehensive marine monitoring and security infrastructure. Thus, the system will create a model that encourages the effective use of digital transformation tools in the transition to sustainable fishing.

The proposed traceability and proactive auditing approach is structurally aligned with SDG 14's objectives of sustainable fishing, combating illegal fishing, and protecting marine ecosystems. The immutable records and auditable event streams provided by the study support evidence-based decision-making processes in port and market inspections, increasing transparency and contributing to food security. Thus, the model enhances policy implementation and provides a governance framework that is consistent with the objectives of the blue economy.

This blockchain-based system offers a strong digital solution in critical areas such as data security, transparency, and system integrity in the fight against IUU fishing and signals that it can be ready for large-scale use with more advanced integrations in the future. This study strengthens the role of digital technologies in marine resource management and proves the digitally supported applicability of sustainable fishing policies.

A limitation of this study is that the evaluations were conducted on a Hardhat local test network; therefore, energy consumption and carbon footprint may not fully reflect production conditions. Future studies could systematically measure per-transaction energy/carbon metrics (kgCO_2e/tx) by piloting on PoS-based networks or L2 solutions and optimize gas consumption through batch processing and off-chain pre-validation. Furthermore, comparative analyses could be planned to balance the additional energy cost impact of privacy and anonymization mechanisms with data minimization.

REFERENCES

Abou Jaoude, J., & George Saade, R. (2019). Blockchain Applications – Usage in Different Domains. *IEEE Access : Practical Innovations, Open Solutions, 7*, 45360–45381. DOI: 10.1109/ACCESS.2019.2902501

Barbosa-Filho, M. L. V., de Souza, G. B. G., de Faria Lopes, S., Hauser-Davis, R. A., Siciliano, S., & da Silva Mourão, J. (2020). Artisanal Fisher Knowledge and Attitudes Concerning Compressor Fishing in a North-Eastern Brazilian Marine Protected Area. *Human Ecology: an Interdisciplinary Journal, 48*(3), 357–366. DOI: 10.1007/s10745-020-00156-2

Bethel, L., Jessen, H., & Hollander, J. (2021). Implementing the Port State Measures Agreement to combat illegal, unreported and unregulated fishing in the Caribbean. *Marine Policy, 132*, 104643. DOI: 10.1016/j.marpol.2021.104643

Brown, S., Katz, D., Korotovskikh, D., & Kullman, S. (2024). Detecting Illegal, Unreported, and Unregulated Fishing through AIS Data and Machine Learning Approaches. *2024 Systems and Information Engineering Design Symposium (SIEDS)*, 319–324. DOI: 10.1109/SIEDS61124.2024.10534704

Busilacchi, S., Butler, J. R. A., Rochester, W. A., & Posu, J. (2018). Drivers of illegal livelihoods in remote transboundary regions: The case of the Trans-Fly region of Papua New Guinea. *Ecology and Society, 23*(1), art46. DOI: 10.5751/ES-09817-230146

Chuaysi, B., & Kiattisin, S. (2020a). Fishing Vessels Behavior Identification for Combating IUU Fishing: Enable Traceability at Sea. *Wireless Personal Communications, 115*(4), 2971–2993. DOI: 10.1007/s11277-020-07200-w

Chuaysi, B., & Kiattisin, S. (2020b). Fishing Vessels Behavior Identification for Combating IUU Fishing: Enable Traceability at Sea. *Wireless Personal Communications, 115*(4), 2971–2993. DOI: 10.1007/s11277-020-07200-w

Collins, M. (2008). Aspects of illegal, unreported and unregulated fishing the Southern Ocean - Rachel J. Baird Springer, Berlin, 2006 ISBN 978-1-4020-5338-2, 309 pages, £65.50. *Antarctic Science, 20*(3), 309–309. DOI: 10.1017/S0954102008001223

Doherty, P. D., Atsango, B. C., Ngassiki, G., Ngouembe, A., Bréheret, N., Chauvet, E., Godley, B. J., Machin, L., Moundzoho, B. D., Parnell, R. J., & Metcalfe, K. (2021). Threats of illegal, unregulated, and unreported fishing to biodiversity and food security in the Republic of the Congo. *Conservation Biology, 35*(5), 1463–1472. DOI: 10.1111/cobi.13723 PMID: 33615559

Drammeh, O. (2000). *Illegal.* Unreported And Unregulated Fishing In Small-Scale Marine And Inland Capture Fisheries.

Food and agriculture organization. (2001). International plan of action to prevent, deter and eliminate illegal, unreported and unregulated fishing (2001). *Journal of International Wildlife Law & Policy, 4*(2), 185–201. DOI: 10.1080/13880290109353986

Gaspar, P., Henriques, V., Fonseca, P., Los, H., Cloarec, M. G., Da Silva, A. J., Da Silva, R. C. A. F., Grosso, N., & Campos, A. (2021). Monitoring Pelagic Fishing Activity in the Northeast Atlantic. *2021 IEEE International Geoscience and Remote Sensing Symposium IGARSS*, 5712–5715. DOI: 10.1109/IGARSS47720.2021.9553509

Hagan, S. (2014). Too Big to Tackle? the Persistent Problem of Pirate Fishing and the New Focus on Port State Measures. *Suffolk Transnational Law Review*, *37*(109).

Han, F., Liu, Y., Tian, H., Li, J., & Tian, Y. (2025). A comprehensive framework incorporating deep learning for analyzing fishing vessel activity using automatic identification system data. *ICES Journal of Marine Science*, *82*(2), fsae166. Advance online publication. DOI: 10.1093/icesjms/fsae166

Kim, D. E., & Lim, S. S. (2024). Economic impacts of the European Union carding system on global fish trade. *Marine Policy*, *165*, 106208. DOI: 10.1016/j.marpol.2024.106208

Kim, Y. H., Howlett, E., Hwang, J. H., & Park, Y. G. (2014). Use of ship traffic density from Automatic Identification System (AIS) and Vessel Monitoring Systems (VMS) in marine spatial planning: A case study in New England coast, USA. *KMI International Journal of Maritime Affairs and Fisheries*, *6*(1), 39–52. DOI: 10.54007/ijmaf.2014.6.1.39

Lee, S., Telesetsky, A., & Schofield, C. (2014). 4 Slipping the Net: Why Is It So Difficult to Crack Down on IUU Fishing? In *Freedom of Navigation and Globalization* (pp. 88–119). Brill | Nijhoff. DOI: 10.1163/9789004284081_007

Lestari, D. I., Putra, A. R., & Larasuci, A. Y. (2020). The main consequences of continued illegal, unreported, and unregulated (IUU) fishing within Indonesian waters for maritime security actors and coastal communities. *Research. Social Development*, *9*(1), e24911566. DOI: 10.33448/rsd-v9i1.1566

Longépé, N., Hajduch, G., Ardianto, R., de Joux, R., Nhunfat, B., Marzuki, M. I., Fablet, R., Hermawan, I., Germain, O., Subki, B. A., Farhan, R., Muttaqin, A. D., & Gaspar, P. (2018). Completing fishing monitoring with spaceborne Vessel Detection System (VDS) and Automatic Identification System (AIS) to assess illegal fishing in Indonesia. *Marine Pollution Bulletin, 131*, 33–39. DOI: 10.1016/j.marpolbul.2017.10.016 PMID: 29106935

Longo, C. S., Buckley, L., Good, S. D., Gorham, T. M., Koerner, L., Lees, S., Liow, S. Y., Oloruntuyi, O., Schley, D., Rice, J., & Currey, R. J. C. (2021). A Perspective on the Role of Eco-Certification in Eliminating Illegal, Unreported and Unregulated Fishing. *Frontiers in Ecology and Evolution, 9*, 637228. Advance online publication. DOI: 10.3389/fevo.2021.637228

Ndiaye, T. M. (2011). Illegal, Unreported and Unregulated Fishing: Responses in General and in West Africa. *Chinese Journal of International Law, 10*(2), 373–405. DOI: 10.1093/chinesejil/jmr012

Palma-Robles, M. A. (2009). *Combating IUU fishing: international legal developments.*

Pratiwi, R. A. I., Fani, L. A., & Kusasi, F. (2024). Blockchain Technology in Fisheries Industry: A Systematic Literature Review. *BIO Web of Conferences, 134*, 05004. DOI: 10.1051/bioconf/202413405004

Shepperson, J. L., Hintzen, N. T., Szostek, C. L., Bell, E., Murray, L. G., & Kaiser, M. J. (2018). A comparison of VMS and AIS data: The effect of data coverage and vessel position recording frequency on estimates of fishing footprints. *ICES Journal of Marine Science, 75*(3), 988–998. DOI: 10.1093/icesjms/fsx230

Sodik, D. M. (2008). Non-legally binding international fisheries instruments and measures to combat illegal, unreported and unregulated fishing. *Australian International Law Journal, 15*(129).

Suherman, A., Santosa, M., Wijayanto, D., Sunoko, R., & Juwana, S. (2020). The Eradication of IUU Fishing in Indonesia for Economic Fisheries. *Saintek Perikanan : Indonesian Journal of Fisheries Science and Technology, 16*, 154–164.

Tetreault, B. J. (2005). Use of the Automatic Identification System (AIS) for Maritime Domain Awareness (MDA). *Proceedings of OCEANS 2005 MTS/IEEE*, 1–5. DOI: 10.1109/OCEANS.2005.1639983

Van Phuong, T., & Pomeroy, R. S. (2023). Addressing Illegal, Unreported and Unregulated Fishing of Vietnamese Fishing Vessels in Foreign Waters. *Asian Fisheries Science, 36*(1). Advance online publication. DOI: 10.33997/j.afs.2023.36.1.003

Welch, C., Paavilainen-Mäntymäki, E., Piekkari, R., & Plakoyiannaki, E. (2022). Reconciling theory and context: How the case study can set a new agenda for international business research. *Journal of International Business Studies*, *53*(1), 4–26. DOI: 10.1057/s41267-021-00484-5

KEY TERMS AND DEFINITIONS

Illegal, Unreported, and Unregulated (IUU) Fishing: Fishing activities conducted in violation of regulations, not reported, or carried out outside the regulatory framework; these practices negatively affect biodiversity, ecosystem balance, and food security.

Automatic Identification System (AIS): A maritime safety system that broadcasts vessels' identity and position information; because signals can be switched off, it has limitations for public oversight and transparency.

Vessel Monitoring System (VMS): A satellite-based system used by competent authorities to monitor fleet movements; public accessibility is limited due to authorization-based access to data.

Smart Contract: A program that automatically enforces predefined rules on-chain; in cases of non-compliance it can halt a transaction (revert) and generate an event log.

Blockchain: A data infrastructure that records transactions across distributed nodes in an immutable and verifiable manner; it enables transparency and auditability.

Web3.py: A software library that enables Python applications to interact with Ethereum-compatible networks.

Ethereum Network: A blockchain platform with a broad ecosystem that supports the execution of smart contracts.

Chapter 4
Cybersecurity in the Water Economy:
Threats and Policies for Critical Infrastructure

Mustafa Bilgehan Imamoğlu
https://orcid.org/0000-0002-3496-2959
Karadeniz Technical University, Turkey

ABSTRACT

The digitalization of water infrastructure enhances operational efficiency but also increases exposure to cyber threats. As Critical Infrastructure essential to national security, public health, and economic stability, water systems are increasingly evolving into complex Cyber-Physical Systems (CPS) with new vulnerabilities. This chapter examines these cybersecurity challenges from a multidisciplinary perspective, analyzing historical and contemporary cyberattacks on SCADA systems to highlight their real-world impact. It further explores the economic implications of cyber risks, including effects on insurance, public financing, and investment decisions. Using a specialized dataset simulating water treatment operations, the performance of AI-based detection models—Random Forest, Support Vector Machines, and Deep Neural Networks—is evaluated, demonstrating their potential to strengthen cyber resilience. Based on these insights, the chapter proposes comprehensive policy and technical strategies to ensure the long-term security of water infrastructure.

DOI: 10.4018/979-8-3373-1777-9.ch004

INTRODUCTION: DIGITALIZING WATER INFRASTRUCTURE AND EMERGING RISKS

In the 21st century, strategic resources such as water, energy, and information have become central tenets of global security policies. Given the pressures of rapidly increasing population, intensive urbanization, and climate change, the entire lifecycle of water, including production, purification, distribution, and reuse, has grown increasingly dependent on sophisticated, high-technology infrastructure systems. This extensive transformation has ushered in the concept of smart water management, where sensor networks, remote monitoring systems, the Internet of Things (IoT), and cloud computing-based control architectures collectively form the digital backbone of the modern water economy (UN-Water, 2022).

One of the most critical components of this digitalization is the widespread adoption of Industrial Control Systems (ICS). Particularly, Supervisory Control and Data Acquisition (SCADA) systems used in water treatment and distribution facilities have been transformed into autonomous structures that collect data from physical processes and can make decisions without continuous human intervention. While these systems provide efficient resource management and cost optimization, they also reveal significant new threat surfaces in terms of cybersecurity (Luiijf & Paske, 2015; TXOne Networks, 2025). The geographical distribution of SCADA devices, their increased remote accessibility through internet connections, and their integration with legacy systems make them highly attractive targets for malicious actors.

Water infrastructure not only carries economic value but is also an indispensable vital service infrastructure for the sustainability of life. Therefore, it is evaluated in the Critical Infrastructure (CI) category along with energy, transportation, and health systems (European Commission, 2023; CISA, 2025). Critical Infrastructure is defined as fundamental systems that ensure the continuity of social order, and any disruption that may occur in these systems can cause serious economic, environmental, or humanitarian losses. In this context, ensuring the security of water infrastructure has become a vital necessity in terms of national security and public policy, going beyond being merely a technical engineering problem.

This chapter adopts an interdisciplinary approach to evaluate the multifaceted cybersecurity problem in water economics with its technical, economic, and managerial dimensions. The study presents a framework through fundamental academic debates and current case analyses in the literature by examining the protection mechanisms at Operational Technology (OT) and Information Technology (IT) levels. In this context, the chapter primarily aims to explain why water infrastructure has increasingly become a target of cyber attacks (1) and to reveal the potential financial impacts of attacks targeting SCADA and IoT components on water economics (2).

Subsequently, it analyzes how these risks can be accurately assessed through public financing mechanisms and insurance systems (3) and the contribution of Artificial Intelligence-Based Intrusion Detection Systems (AI-IDS) in enhancing the cyber resilience of this infrastructure (4). These fundamental research questions will be addressed comprehensively in the subsequent sections of the study.

1. WATER SYSTEMS AS CRITICAL INFRASTRUCTURE

In the last quarter of the 20th century, with the deep integration of Information and Communication Technologies into fundamental economic sectors, the concept of Critical Infrastructures (CI) underwent a radical transformation. The United States' Presidential Decision Directive No. 63 of 1998 officially designated the energy, transportation, water, finance, and telecommunications sectors as "fundamental pillars of national security," making the protection of these systems a strategic priority (Lewis, 2006; GAO, 2024). Similarly, the European Union established a similar framework with the European Programme for Critical Infrastructure Protection (EPCIP) Directive 2008/114/EC. Furthermore, the NIS2 Directive, which entered into force in 2022, placed water supply at the center of cybersecurity policies, classifying this sector among the most critical areas (European Commission, 2023).

Most importantly, Critical Infrastructures (CIs) are not merely physical systems; rather, they form a whole with Cyber-Physical Systems (CPS) that provide digital control of these systems. In this context, when physical production and distribution systems, such as water infrastructure, are integrated with advanced information technology (IT) systems, highly complex structures emerge that require protection at both the physical and digital levels. A fundamental characteristic noteworthy in global infrastructure systems is the increasingly interconnected dependency structure. Particularly, this relationship, often referred to in the literature as the "water-energy-information nexus," plays a key role in Critical Infrastructure security efforts.

The nature of this nexus can be broken down into two primary dependencies:

Water–Energy Interdependence: Energy production processes are inherently dependent on water resources. Similarly, the treatment and distribution of water exhibit a notable reliance on energy. For example, water is required for the operation of turbine systems in hydroelectric power plants as well as for maintaining cooling cycles in thermal power plants. In addition, the functionality of operational equipment used in a water treatment facility—such as pumps, valves, and chemical dosing systems—is directly supported by electrical energy. This resulting mutual dependency indicates that an interruption in the energy infrastructure can directly affect the operational continuity of the water infrastructure, while a failure in the water supply can likewise impose immediate constraints on energy production capacity. For this reason, the interconnected structure of water and energy systems is of critical importance in terms of sustainability and security (Olsson & Lund, 2017).

Information-Water Interaction: Digital sensors, Internet of Things (IoT) devices, SCADA networks, and cloud-based monitoring platforms have enabled the implementation of data-driven decision support systems in water infrastructure. While this technological transformation provides significant contributions to the more efficient, resilient, and sustainable management of water systems, it simultaneously creates the foundation for the emergence of new threat vectors in the cybersecurity domain. Therefore, both technological innovations and security measures need to be addressed in an integrated manner.

Consequently, Critical Infrastructure (CI) security should not remain limited to targeting the protection of isolated systems but should evolve into an approach aimed at ensuring the holistic security of the interdependent networks of these systems. For instance, a cyber attack targeting water infrastructure can not only cause disruption of water services but also produce cascading systemic consequences that seriously affect the energy supply chain, health infrastructure, and indirectly public safety.

2. CYBER THREATS TO WATER INFRASTRUCTURE AND CASE ANALYSES

SCADA systems used in water infrastructure generally rely on a hierarchical architecture consisting of three fundamental layers in which specific functional boundaries are defined between each layer, as visually illustrated in Figure 1. However, the inability to maintain these boundaries consistently in practice seriously jeopardizes the holistic security of the system. Particularly, insufficient network segmentation between internet-exposed remote access points and corporate IT and OT networks constitutes priority target areas for attackers.

The fundamental security vulnerabilities in this architecture stem from the combination of technological legacy and the human factor. Industrial communication protocols such as Modbus and DNP3 were not designed to have integrated authentication or encryption capabilities. Therefore, these protocols can be easily exploited in today's complex threat environment (Zetter, 2014). Furthermore, the components of SCADA systems typically have a long operational lifespan of 15-20 years, and this situation leads to the continued use of legacy operating systems. For instance, unsupported systems such as Windows 7 are still actively utilized. The inability of these legacy systems to receive necessary security patches makes them extremely vulnerable to cyber attacks (CISA, 2021a; Bello et al., 2023).

In addition, an inadequate separation between the IT and OT networks facilitates lateral movement, allowing an attacker who successfully infiltrates the IT network to pivot into the OT domain. This scenario generates cascading risks that can extend to the remote manipulation of physical production processes. Finally, the human element frequently represents one of the weakest links in system security. The use of weak passwords, successful phishing campaigns, and social engineering methodologies function as the initial compromise vector in the majority of cyber attacks targeting water infrastructure. Considering all these elements collectively, it is evident that technical vulnerabilities in SCADA architectures are shaped not only at the hardware or software levels but also by organizational security culture and human behavior. Therefore, ensuring cybersecurity in water infrastructure necessitates a comprehensive security management approach that extends far beyond mere technical countermeasures.

Figure 1. Typical Water Treatment Plant SCADA Architecture and Threat Surfaces.

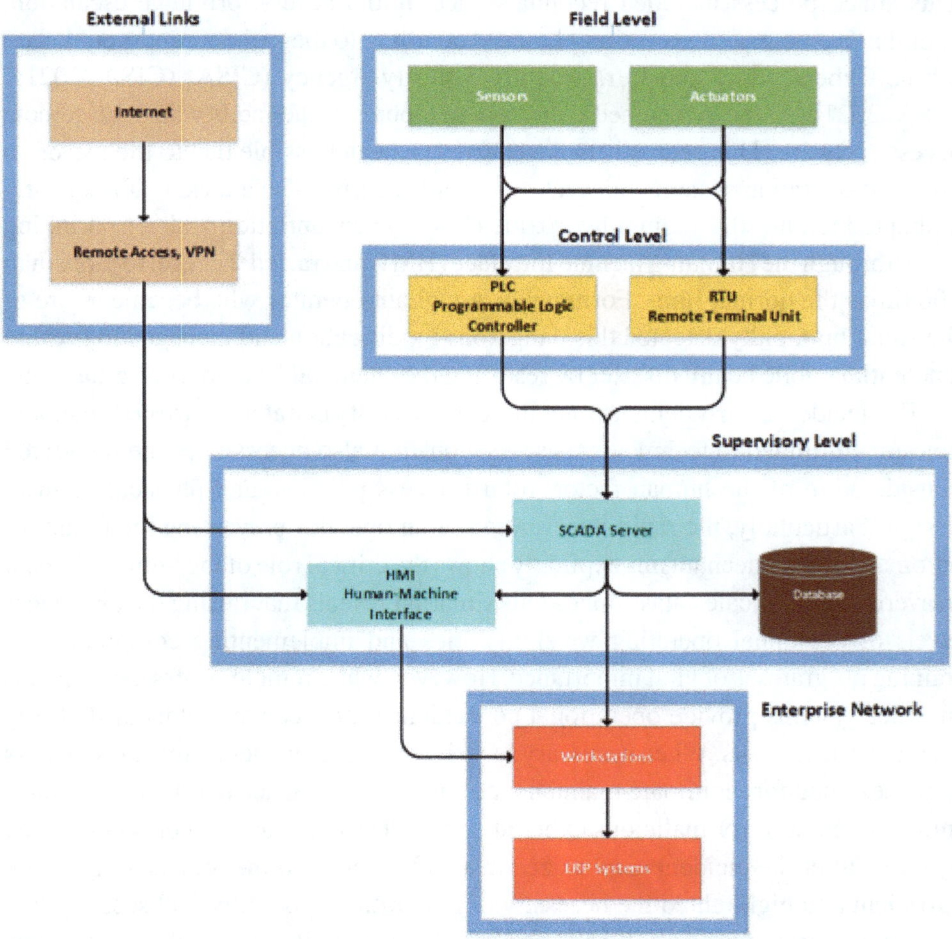

The incident that occurred in the Maroochy Shire water and sewage system in Queensland, Australia, in 2000 stands as one of the first documented cyber attacks targeting SCADA systems. A former employee of the contracting firm that installed the facility's SCADA system took control of 46 separate pump stations using a laptop and a radio transmitter. Over a two-month period, the attacker repeatedly compromised the system, leading to the discharge of approximately 800,000 liters of untreated sewage into local parks, rivers, and a hotel garden (Abrams & Weiss, 2008). This case starkly highlighted the destructive potential of insider threats and deficiencies in access management controls. The attacker successfully evaded detection for an extended period due to their deep familiarity with the system's operational parameters.

On February 5, 2021, the incident at the Oldsmar water treatment facility in Florida, USA, strikingly revealed the risks brought by modern remote access technologies. This attack process included reconnaissance, initial access, privilege escalation, lateral movement, and execution phases. According to the official report published by the Cybersecurity and Infrastructure Security Agency (CISA) (CISA, 2021a; CISA, 2021b], attackers gained entry to a computer at the facility through remote access software. This access is believed to have been possible due to the use of an outdated operating system and weak password security. The attacker subsequently attempted to alter the sodium hydroxide (NaOH) concentration added to drinking water through the Human-Machine Interface (HMI) and raised the level to more than 100 times the normal limit. Fortunately, a vigilant operator who became aware of the situation rapidly detected this dangerous modification and managed to prevent a potential public health disaster by restoring the chemical level to its original state.

This incident clearly demonstrates that cybersecurity in water infrastructure should not remain limited to technical measures alone but also necessitates the integrated consideration of the human factor, robust access policies, and physical security design. Particularly, the rapid intervention of an operator preventing the failure of automatic alarm mechanisms explicitly shows the critical role of the human element in averting a full-scale catastrophe. This situation reveals that raising the awareness levels of personnel operating water facilities and implementing comprehensive training programs is of vital importance. However, while remote access mechanisms in water systems provide operational convenience, they can simultaneously bring serious security risks. When necessary security protocols are neglected, these access tools designed for legitimate maintenance and monitoring activities can transform into an open door for malicious cyber attackers. Technical assessments conducted by CISA after the incident emphasized that software-based measures alone are not sufficient and highlighted the necessity of integrating cyber-physical security layers into the design process. It has been noted that even if control systems become dysfunctional, the correct design of physical components such as pump capacities, chemical dosing limits, and tank volumes is critically important to prevent dangerous mixing of chemicals. These findings strongly reinforce that cybersecurity in water infrastructure is a multi-layered structure and that technical, operational, and human-centered measures must be integrated in a harmonious manner.

3. ECONOMIC DIMENSIONS OF RISK AND POLICY ANALYSIS

While cybersecurity in water infrastructure is often perceived primarily as a cost burden, it should, in fact, be strategically viewed as a preventative investment that safeguards economic sustainability. To evaluate the economic rationality of these cybersecurity expenditures, a Cost-Benefit Analysis (CBA) framework proves highly suitable.

The Costs encompass direct expenditure, such as those for security software, hardware, personnel training, audits, and cyber insurance premiums. Conversely, the Benefits are defined as the potential losses avoided by successfully preventing a cyber attack. These avoided losses can be modeled using the Annualized Loss Expectancy (ALE) formula:

$$ALE = Single\ Loss\ Expectancy\ (SLE) * Annualized\ Rate\ of\ Occurrence\ (ARO)$$

Within the scope of cyber risk analysis, the Single Loss Expectancy (SLE) is one of the most fundamental metrics used to assess the potential economic impact of an attack. The SLE represents the total financial loss likely to materialize should a specific cyber incident occur. Critically, this loss includes not only direct costs such as system repair, service interruption, and liability payments but also indirect costs, including reputational damage, legal sanctions, and public health expenditures.

The other pivotal variable is the Annualized Rate of Occurrence (ARO), which defines the estimated probability of a specific type of attack occurring within a one-year period. This rate is utilized as a core parameter in calculating ALE.

In this established framework, the Net Present Value (NPV) approach is applied to evaluate the economic soundness of a security investment. The NPV is calculated by subtracting the investment cost from the total benefit the investment provides. An investment is considered economically rational and sustainable if the NPV yields a positive value (Şentürk et al., 2016). However, for sectors that serve the public interest, such as water infrastructure, this analysis must not be confined solely to financial indicators. It is imperative that positive externalities specifically impact on public health, environmental sustainability, and societal welfare are integrated into the NPV calculation. Incorporating these externalities ensures that public investments are evaluated within a framework focused on social benefit as well as financial feasibility.

4. METHODOLOGY

This section details the development of an Artificial Intelligence-based Intrusion Detection System (AI-IDS) designed specifically for the detection of cyber attacks targeting SCADA systems within water infrastructure. The application was performed using a synthetic dataset meticulously constructed based on the characteristics of the widely recognized SWaT (Secure Water Treatment) dataset (Goh et al., 2017). It should be noted that SWaT is a physical CPS testbed that accurately simulates the six-stage process of a real-world water treatment facility. Consequently, this environment encompasses the entire water purification cycle, ranging from the initial raw water intake (Stage 1) to the final backwash procedures (Stage 6).

4.1. Dataset Characteristics and Attack Scenario Design

The operational activities of the simulated water treatment plant spanned an 11-day period. The initial seven days represent the system's normal operating conditions and comprise 7,000 instances, recorded at one-minute intervals. The subsequent four days cover periods during which various attack scenarios were implemented, consisting of 4,000 instances. The total size of the dataset, with its class distribution presented in Figure 2, reaches 11,000 instances. Of this total, 7,000 instances (63.6%) represent normal operational conditions, while 4,000 instances (36.4%) correspond to various cyber attack scenarios. Although this distribution reflects the real-world reality that normal operation is observed far more frequently than attack states in SCADA systems, it presents a balanced structure that avoids severe class imbalance issues. Using a stratified sampling technique, the dataset was partitioned into a 70% training subset (7,700 instances: 4,900 normal, 2,800 attack) and a 30% testing subset (3,300 instances: 2,100 normal, 1,200 attack), thereby ensuring that the original class proportions were accurately maintained in both subsets. This carefully balanced distribution allows the machine learning models to effectively learn both normal and attack behaviors with sufficient examples, minimizing learning problems often caused by class imbalance (e.g., over-prediction of the majority class). Furthermore, the 36.4% proportion of attack instances satisfies the "minimum 30% minority class" criterion often recommended in the literature, supporting reliable performance from the models in attack detection.

Figure 2. (Left) Percentage distribution of Normal and Attack samples in the total dataset. (Right) Distribution of sample numbers in the training and test sets.

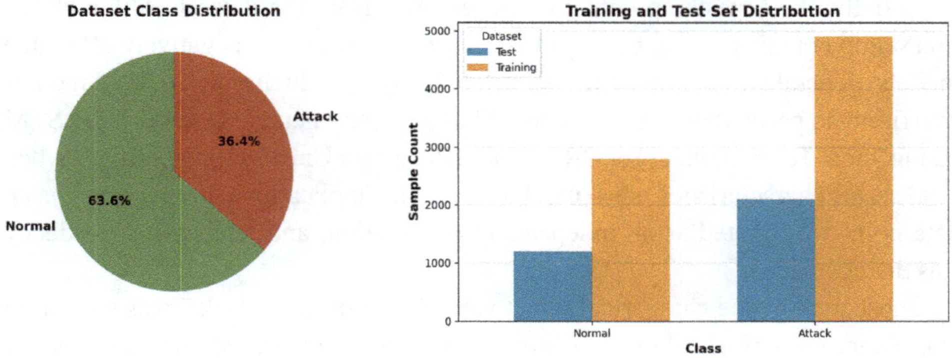

Throughout the data collection process, the sampling frequency was fixed at once per minute by taking measurements every minute. The dataset consists of a total of 25 features derived from sensor and actuator measurements corresponding to six different operational stages of the water treatment facility. This comprehensive structure enables the detailed observation of interactions between both the physical and digital components of the system.

The attack scenarios implemented in the synthetic dataset were designed to represent the most common manipulation techniques observed in SCADA systems, inspired by real-world incidents. Particularly, the pH Manipulation Attack, which directly references the Oldsmar incident, simulates the addition of excessive sodium hydroxide (NaOH) to drinking water by raising AIT201 (pH sensor) values from the normal level of 7.0 to 11.5. Another scenario, the Flow Anomaly Attack, models the compromise of pump control by increasing FIT101 (flow sensor) values from 2.5 m^3/hr to 5.0 m^3/hr. Similarly, the Level Manipulation scenario mimics a tank draining attack by decreasing LIT101 (tank level) values from 500 mm to 200 mm, while the Pressure Attack simulates excessive pressure application that could damage ultrafiltration membranes by raising DPIT301 (differential pressure) values from 20 psi to 50 psi.

4.2. Data Preprocessing Steps

A systematic preprocessing pipeline was applied to the dataset to ensure the machine learning models could operate effectively. This process consisted of five essential steps designed to transform the raw SCADA data into a format suitable for model training.

Initially, the timestamp column was removed from the dataset. Although the timestamp contains sequential information, its inclusion can negatively affect the models' generalization capability, specifically by introducing the risk of models overfitting to particular time segments. While recurrent neural networks like LSTM (Long Short-Term Memory) or GRU (Gated Recurrent Unit) would be utilized when time-series characteristics are critical in real-time applications, in this study, each data point was treated as an independent observation, and temporal dependency was disregarded.

Subsequently, the categorical target variable, "Normal/Attack," was converted into binary numerical values. This process, known as label encoding, is necessary for machine learning algorithms to process categorical data. The normal operation state was encoded as 0, and the attack state was encoded as 1. This binary coding simplifies the mathematical formulation of the classification problem and permits the use of sigmoid or softmax activation functions in the models' output layers.

The third step involved separating the dataset into independent variables (X) and the dependent variable (y). The matrix of X comprises the 25 sensor and actuator measurements, which represent the physical parameters collected from the six distinct operational stages of the water treatment facility (raw water intake, chemical dosing, ultrafiltration, dechlorination, reverse osmosis, and backwash). The y is the binary label vector indicating whether each observation belongs to the normal or attack class.

In the fourth step, the dataset was carefully divided into training and test subsets using the stratified sampling method. Stratified sampling has critical importance as it ensures that the class distribution in both subsets is preserved at the same proportions as the original dataset. 70% of the dataset was allocated to the training set (7,700 samples: 4,900 normal, 2,800 attack) and 30% to the test set (3,300 samples: 2,100 normal, 1,200 attack). This ratio is a widely accepted standard in the machine learning field and both provides a sufficient number of training samples and enables reliable measurement of test performance. Particularly in imbalanced datasets, stratified sampling plays a vital role; because random splitting can increase the risk of insufficient representation of the minority class (the attack class in this study) in the test set.

The fifth and final step encompasses the normalization of features. Different sensors produce data at different scales and units; for instance, pH values can vary between 0 and 14, while pressure values range between 0 and 100 psi, and flow values range between 0 and 10 cubic meters/hour. These scale differences can create serious challenges, especially for models using distance-based algorithms or gradient-based optimization techniques. Therefore, all features were standardized using StandardScaler. This process balanced the impact of features at different scales on model performance by bringing the mean of all features to 0 and the standard deviation to 1. As an important point, normalization was performed using the mean and standard deviation values learned from the training set, and these values were also applied to the test set. This method prevented data leakage, enabling a more accurate evaluation of the model's performance under real-world conditions.

4.3. Methodological Framework

The methodological flow of this study, systematically summarized in Figure 3, is based on a carefully designed and reproducible structure. In the first phase, detailed and comprehensive data preparation processes were implemented to transform raw SCADA data into a format compatible with machine learning algorithms. In the second phase, model selection was performed and the performances of three algorithms representing different learning paradigms were compared: Random Forest (RF) representing the ensemble learning approach, Support Vector Machine (SVM) using kernel-based classification, and Deep Neural Network (DNN) representing deep learning. This diversified algorithmic structure provides the opportunity to analyze in depth the suitability of different approaches to a complex problem such as cyber attack detection in water infrastructures.

Figure 3. Data Preprocessing and Methodology Flowchart

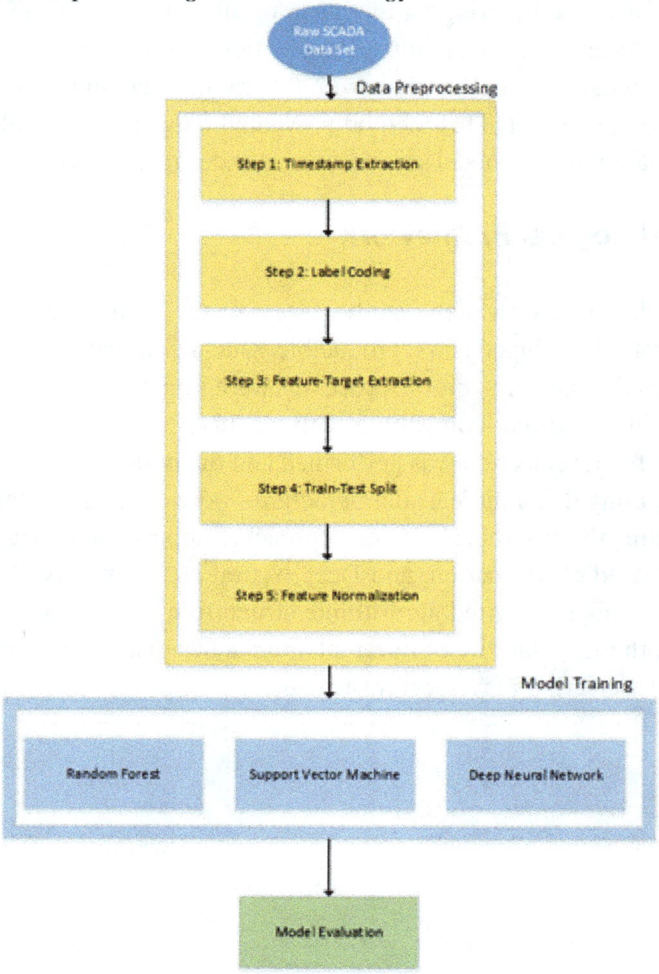

The third phase comprises the training procedure of the models. Each model was subjected to a specific learning process using the corresponding training algorithm. To ensure optimal performance, the hyperparameters were carefully selected based on findings obtained from the literature and preliminary experimental evaluations. These hyperparameters included the kernel type for RF (Radial Basis Function), the number of trees (n_estimators = 100), and the maximum depth (max_depth = 20); the regularization parameter for SVM (C = 1.0); and, for the DNN, the number of hidden layers (hidden_layer_sizes = (128, 64, 32)) together with the activation function (relu).

The fourth phase focuses on the model evaluation procedure. In this stage, the trained models generated predictions on an unseen test set, and their performances were comprehensively analyzed through five fundamental metrics: accuracy, precision, recall (sensitivity), F1-score, and ROC-AUC. The multi-metric evaluation approach enabled a detailed examination of the models from different perspectives, allowing a more balanced, comprehensive, and objective comparison of their overall performances.

In the fifth phase, the cross-validation technique was employed. The generalization capability of the models was assessed using the 5-Fold Cross-Validation (5-Fold CV) technique. In this approach, the training dataset was partitioned into five equal subsets, and each model was trained five times, with a different subset used as the validation set during each iteration. The cross-validation results provided insights into whether the models exhibited consistent performance not only on a single test set but also across various data partitions.

In the sixth phase, a comprehensive error analysis was conducted. Misclassified instances were examined in detail, and the proportions of false positive and false negative error types were calculated with great care. This analysis not only identified the types of errors to which the models are more susceptible but also provided a clearer basis for evaluating potential risks that may arise when the models are deployed in continuous integration (CI) environments. Furthermore, this process plays a critical role in planning strategic improvements aimed at enhancing the overall performance of the models.

The final phase entailed a complexity analysis. The models were assessed based on practical metrics such as computational cost, training time, prediction speed, and model size. This analysis ensures that the evaluation considers not only the theoretical performance of the models but also their practical deployability in real-time SCADA systems. Metrics like the number of instances that can be processed per second and the model size are of paramount importance in industrial environments with limited computational resources. This seven-stage methodological framework underpins the scientific rigor and reliability of the study's findings while simultaneously providing a reproducible roadmap for future research.

4.4. Model Performance Results and Evaluation

Table 1 presents a summary of the performance metrics of the trained models on the test dataset, while Figure 4 provides a visual comparison of these metrics. An objective assessment of the models' expected real-world performance is ensured by the test set, which contains 3,300 instances not encountered by the models during the training phase.

Table 1. Model performance metrics

Model	Accuracy	Precision	Recall	F1-Score	ROC-AUC
RF	0.9142	0.9011	0.9719	0.9352	0.9584
SVM	0.9170	0.9047	0.9719	0.9371	0.9593
DNN	0.9203	0.9154	0.9638	0.9390	0.9613

A complete analysis of the performance metrics is vital for comprehending the advantages and disadvantages associated with the models under investigation. According to the Accuracy metric, which represents the ratio of correct predictions across all predictions, the Deep Neural Network (DNN) model achieves the optimum accuracy of 92.03%. Based on this finding, it is evident that the model successfully classifies both normal instances and attack cases within the overall context. The Support Vector Machine (SVM) model attains an accuracy level of 91.70%, while the Random Forest (RF) model reaches an accuracy level of 91.42%. These results indicate that both models achieve a comparable level of performance. The marginal difference in accuracy, approximately 1%, among the three models suggests that all three possess an acceptable capacity for detecting cyber attacks within water infrastructure systems.

Figure 4. Comparative results of three models on five different performance metrics.

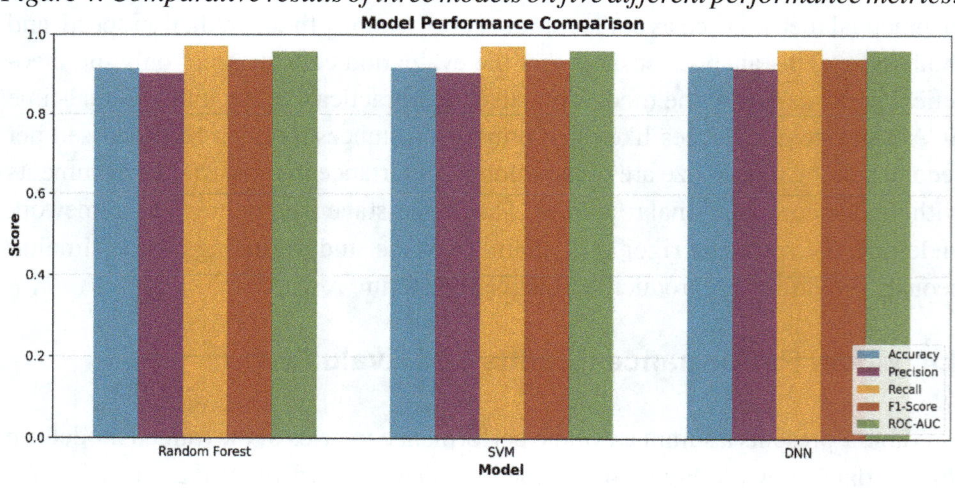

The Precision metric measures the proportion of instances predicted as attacks that were truly attacks. A high precision value signifies a low False Positive (false alarm) rate for the model. In CI environments, false alarms increase the workload on operators and, over time, can lead to alarm fatigue, thereby escalating the risk of genuine threats being overlooked. The DNN model demonstrated the best precision at 91.54%, meaning that 91 out of every 100 instances flagged as an attack by the DNN were actual attacks. The SVM (90.47%) and RF (90.11%) models followed closely behind. These results indicate that all three models are capable of maintaining false alarm rates at reasonable levels.

The Recall (Sensitivity) statistic quantitatively specifies the proportion of actual attack incidents that are correctly detected. A high recall value is indicative of a low rate of False Negatives (missed attacks). In vital systems such as water infrastructure, the False Negative represents the most perilous type of error, as the failure to detect a genuine cyber attack can pose a direct risk to public health and national security. The Random Forest (RF) and Support Vector Machine (SVM) models achieved the highest recall value of 97.19%. This finding confirms that both models effectively identified nearly all attack cases within the test sample. The Deep Neural Network (DNN) model secured the third position with a recall rate of 96.38%. Although the recall of the DNN is approximately 1% lower than the other two models, which suggests a slightly higher rate of missed attacks, a recall value exceeding 96% remains an acceptable standard for Critical Infrastructure (CI).

The F1-Score metric represents the harmonic mean of Precision and Recall, and its optimization aims for a balanced assessment of both metrics. The F1-Score is generally considered the most reliable performance indicator, particularly in imbalanced datasets, as it simultaneously accounts for both the false alarm rate and the missed attack rate. The DNN model achieved the highest F1-Score at 93.90%, indicating it attained the best balance between precision and recall. SVM (93.71%) and RF (93.52%) followed. The minute difference of only 0.38% among the F1-Scores reveals that their performances are remarkably similar. Nonetheless, even this slight edge by the DNN can translate into a meaningful difference when processing thousands of instances in real-time SCADA systems.

The Receiver Operating Characteristic – Area Under the Curve (ROC-AUC) metric is a comprehensive statistic that summarizes a model's performance across all possible classification threshold values. The ROC curve provides a graphical representation of the relationship between the True Positive Rate (TPR) and the False Positive Rate (FPR) at varying threshold levels. The larger the area under the curve (AUC), the more effectively the model can distinguish between different classes. In contrast, a random classifier yields an AUC of 0.5, whereas a perfect classifier achieves an AUC of 1.0.

Figure 5. ROC curves of the models

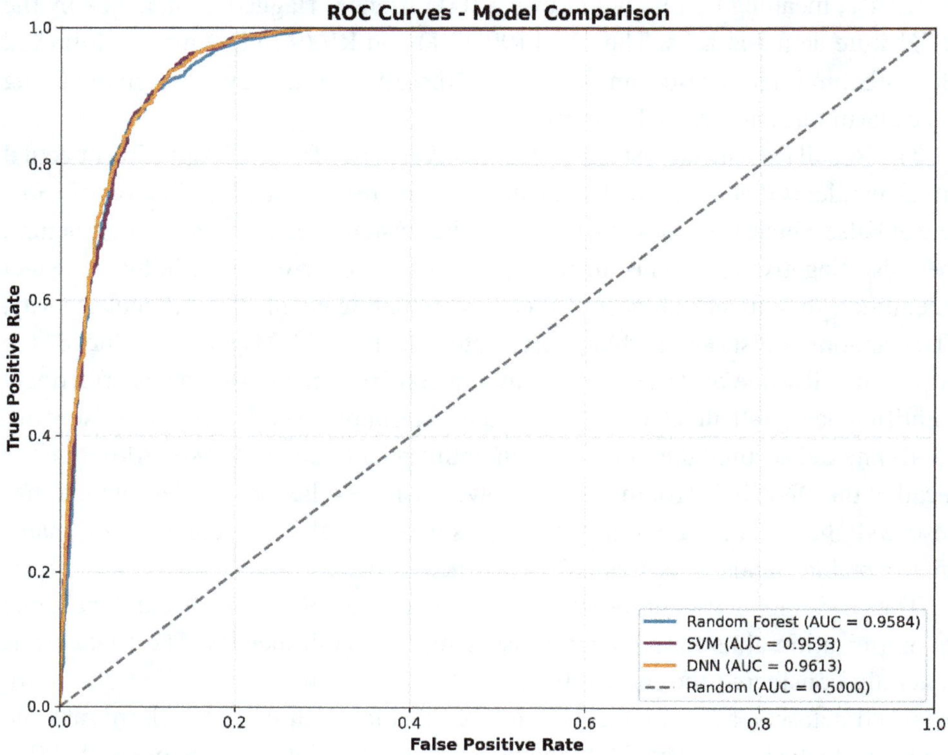

The ROC curves presented in Figure 5 compare the classification performance of the three models by showing the relationship between TPR and FPR at different decision thresholds. All three models demonstrated excellent discriminative ability, with their AUC values (DNN=0.9613, SVM=0.9593, RF=0.9584) being substantially higher than the random prediction line (AUC=0.5). Although the models' performances are very close, DNN model exhibited the best performance, as its ROC curve is closest to the ideal point (0,1) in the upper-left corner. This superiority is particularly noticeable at low FPR values, where the DNN can maintain high attack detection rates (e.g., TPR ≈0.85 when FPR =0.05) even with a very low false alarm rate.

In conclusion, the results demonstrate that all three models achieved high success in detecting cyber attacks within water infrastructure. Specifically, the DNN model showed the most balanced and reliable performance across critical metrics like F1-Score and ROC-AUC. The DNN's advantage stems from its ability to maintain high detection accuracy while simultaneously keeping the false alarm rate low. Conversely, RF and SVM models proved their capability to capture almost all attacks, showing superiority, particularly in the recall metric. These findings suggest that operational priorities should guide model selection: if the priority is minimizing false alarms, DNN is preferable; if the priority is ensuring no attack is missed, RF or SVM may be chosen.

5. CONCLUSION AND POLICY RECOMMENDATIONS

This study has established that the digitalization of water infrastructure inevitably leads to escalating cybersecurity risks, which carry the potential for severe economic and societal consequences. The case analyses demonstrated the vulnerability of both legacy and modern systems to various attack vectors, while the application section provided empirical evidence that AI-IDS possess a high accuracy potential for identifying such threats. In light of these findings, enhancing cyber resilience in the water economy mandates the adoption of a holistic approach where cybersecurity is intrinsically integrated with physical security and operational processes; consequently, the mandatory implementation of cyber-physical mechanisms that physically prevent dangerous operational levels is required. Concurrently, minimum cybersecurity standards similar to those set forth by the NIS2 Directive must be established for water service providers, accompanied by regular independent audits and a clear roadmap for updating legacy systems. Furthermore, the use of AI for anomaly detection should be encouraged through R&D support and data sharing platforms. Equally critical is investing in the human factor through continuous training and exercises for personnel, alongside strengthening public-private partnership mechanisms to facilitate real-time threat intelligence sharing among national response centers, water utilities, and technology providers.

Based on these comprehensive findings, the following concrete and multi-layered policy recommendations are presented for policymakers, regulatory bodies, and sector operators to enhance the cyber resilience of water infrastructure:

- *Development of a Holistic and Risk-Oriented National Strategy:* A cybersecurity strategy specific to the water sector must be built upon risk scenarios that explicitly address cross-sector interdependencies, such as the water-energy nexus. This strategy should encompass not only technical controls but also the dimensions of governance, human capital, and financing. Drawing inspiration from international frameworks like the EU's NIS2 Directive, minimum cybersecurity standards must be established for all water service providers, with compliance verified through regular, independent audits.
- *Strengthening Public-Private Collaboration and Information Sharing:* The real-time exchange of threat intelligence should be actively promoted through

trust-based platforms, such as Water and Wastewater Information Sharing and Analysis Centers. Furthermore, regular cyber exercises involving public and private sector operators will significantly boost incident response capacity and stakeholder coordination. As emphasized by the OECD, these collaborations must be built upon transparent governance models that ensure the fair distribution of risks and responsibilities.

- *Financial Incentives and Capacity Building:* It is critically important to enhance the cybersecurity capacity of small and medium-sized municipal water utilities, which often operate with limited resources. Drawing on models such as the Drinking Water State Revolving Fund (DWSRF) in the United States, financial support including grants, low-interest loans, and free technical consulting should be provided to help these institutions achieve essential cyber hygiene standards.
- *Investment in the Human Factor and Cultivating an Awareness Culture:* Technology alone is insufficient; the weakest link is frequently the human element. Continuous and mandatory training programs must be implemented to raise the cybersecurity awareness of all personnel, from plant operators to executives. These training sessions should include practical applications such as phishing simulations, social engineering tactics, and incident response drills.
- *Supporting R&D and the Domestic Technology Ecosystem:* Developing advanced defensive systems, such as AI-IDS, using domestic capabilities will reduce external dependence and strengthen national cybersecurity capacity. To achieve this, public support and incentives must actively promote R&D collaborations among universities, research centers, and sector operators.

In conclusion, the digital transformation of water infrastructure is an unavoidable process that carries the potential for a silent security crisis. Averting this crisis demands a multilayered approach to resilience that integrates technological investment, rational economic models, effective policy frameworks, and international cooperation. The secure water systems of the future will be built not only with smart technologies but also with a culture of proactive governance, continuous learning, and cross-stakeholder collaboration.

REFERENCES

Abrams, M., & Weiss, J. (2008). *Malicious control system cyber security attack case study – Maroochy Water Services*. NIST Special Publication., https://www.mitre.org/sites/default/files/pdf/08_1145.pdf

Bello, A., Jahan, S., Farid, F., & Ahamed, F. (2023). A systemic review of the cybersecurity challenges in Australian water infrastructure management. *Water (Basel)*, *15*(1), 168. DOI: 10.3390/w15010168

CISA. (2021a). Compromise of U.S. water treatment facility (Alert Code AA21-042A). Cybersecurity and Infrastructure Security Agency. https://www.cisa.gov/news-events/cybersecurity-advisories/aa21-042a

CISA. (2021b). Compromise of U.S. Water Treatment Facility. Retrieved from https://www.cisa.gov/news-events/cybersecurity-advisories/aa21-042a

CISA. (2025). Water and Wastewater Systems Sector. Retrieved from https://www.cisa.gov/topics/critical-infrastructure-security-and-resilience/critical-infrastructure-sectors/water-and-wastewater-sector

European Commission. (2023). Directive (EU) 2022/2555 on measures for a high common level of cybersecurity across the Union (NIS2 Directive). Official Journal of the European Union. https://eur-lex.europa.eu/eli/dir/2022/2555/oj

GAO, Government Accountability Office. (2024). Critical Infrastructure Protection: EPA Urgently Needs a Strategy to Address Cybersecurity Risks to Water Systems. GAO-24-106744. Retrieved from https://www.gao.gov/products/gao-24-106744

Goh, J., Adepu, S., Junejo, K. N., & Mathur, A. (2017). A Dataset to Support Research in the Design of Secure Water Treatment Systems. In Havarneanu, G., Setola, R., Nassopoulos, H., & Wolthusen, S. (Eds.), Lecture Notes in Computer Science: Vol. 10242. *Critical Information Infrastructures Security. CRITIS 2016*. Springer., DOI: 10.1007/978-3-319-71368-7_8

Lewis, T. G. (2006). *Critical infrastructure protection in homeland security: Defending a networked nation*. John Wiley & Sons. DOI: 10.1002/0471789542

Luiijf, H., & Paske, B. (2015). Cyber Security of Industrial Control Systems..

Olsson, G., & Lund, P. (2017). Water and Energy - Interconnections and Conflicts. 1(5), 1700056. https://doi.org/DOI: 10.1002/gch2.201700056

Şentürk, H., Çil, C. Z., & Sağıroğlu, Ş. (2016). Siber Güvenlik Yatırım Kararları Üzerine Literatür İncelemesi. *Politeknik Dergisi*, *19*(1), 39–51.

TXOne Networks. (2025). Cyber Threats to Water and Wastewater Sector. Retrieved from https://www.txone.com/blog/cyber-threats-to-water-and-wastewater-sector/

UN-Water. (2022). *The United Nations world water development report 2022: Groundwater: Making the invisible visible.* UNESCO.

Zetter, K. (2014). *Countdown to zero day: Stuxnet and the launch of the world's first digital weapon.* Crown.

Chapter 5
Inclusive Pathways to Sustainable Marine Enterprises: The Role of Human Resources

Mochamad Mochklas
https://orcid.org/0000-0001-7674-2551
Universitas Muhammadiyah Surabaya, Indonesia

Dwi Songgo Panggayudi
https://orcid.org/0009-0009-6841-4131
Universitas Muhammadiyah Surabaya, Indonesia

Djoko Soelistya
https://orcid.org/0000-0001-8085-2614
Universitas Muhammadiyah Gresik, Indonesia

Sofiah Nur Iradawaty
https://orcid.org/0009-0009-7572-5544
Universitas Yos Soedarso, Indonesia

ABSTRACT

An inclusive approach to creating sustainable entrepreneurship in the marine sector is essential, with human resources as a strategic factor. Amidst increasing environmental challenges, sustainable management of marine resources requires green entrepreneurship. Human resources play a role in driving innovation, strengthening local capacity, and ensuring the involvement of women, youth, and indigenous communities. Policies that support skills education, economic empowerment, and access to green technology are needed. Collaboration between government, the

DOI: 10.4018/979-8-3373-1777-9.ch005

private sector, academia, and local communities will create sustainable business models that support SDG 8 (Decent Work and Economic Growth) and SDG 14 (Marine Ecosystems).

INTRODUCTION

Over the past two decades, coastal regions around the world have increasingly drawn attention as societies confront two competing imperatives: economic development and environmental sustainability. The concept of the *blue economy* has emerged as a guiding paradigm, advocating the responsible use of marine resources to generate prosperity while safeguarding ecosystems from irreversible damage. This issue is particularly pressing in Indonesia, where millions of coastal residents depend on the sea for their livelihoods. Fisheries, aquaculture, and marine tourism remain central to household economies, yet reliance on these sectors also makes communities highly vulnerable. Problems such as overfishing, habitat degradation, and plastic pollution undermine ecological resilience, while limited access to finance, weak institutional arrangements, and uneven technological adoption continue to constrain social and economic progress (Gaines et al., 2018; Hoegh-Guldberg et al., 2019; Issifu et al., 2022).

At the global level, commitments to the *Sustainable Development Goals (SDGs)* have reinforced the urgency of linking coastal development with sustainability principles. Targets related to decent work, climate action, responsible production, and marine ecosystem health highlight that livelihoods and ecological balance must go hand in hand (United Nations, 2020). Despite numerous conservation and livelihood-improvement programs, their effectiveness often remains limited. In many cases, the gap between ambitious policy goals and the lived reality of coastal communities is wide—particularly where local capacities to adapt, innovate, and sustain green enterprises are still underdeveloped (Masni et al., 2024).

Human resources therefore stand at the center of this challenge. Theories of *human capital* and the *capability approach* stress that education, health, and skills are not merely background factors but critical levers that enable individuals to transform available resources into tangible outcomes (Becker, 1993; Sen, 1999). Applied to coastal contexts, this means that entrepreneurship will not be sustainable unless communities are equipped with practical skills, mentoring opportunities, and supportive institutions. Put differently, empowerment is not a supplementary concern—it is the very foundation upon which sustainability can flourish.

Beyond individual capabilities, frameworks such as *green business models* and the *circular economy* emphasize innovation, efficiency, and resource reuse (Boons & Lüdeke-Freund, 2013; Geissdoerfer et al., 2017). While valuable, such approaches often place heavy emphasis on technology and overlook the social and cultural realities of coastal life. In contrast, community-based entrepreneurship highlights local traditions, collective norms, and inclusive leadership—dimensions that resonate strongly with coastal societies, where economic activity is deeply intertwined with social values (Peredo & Chrisman, 2006). Recent empirical studies confirm this: leadership skills, digital literacy, and gender inclusivity increasingly determine whether marine-based enterprises succeed or fail (Fudge et al., 2023; Masni et al., 2024; Mochklas et al., 2024).

Despite growing interest, the scholarship remains fragmented. Fisheries research tends to focus on ecological risks, while studies of coral reefs and marine pollution primarily emphasize environmental degradation. Much of the literature on green entrepreneurship, meanwhile, is dominated by technological narratives, leaving community readiness underexplored. A more comprehensive framework is therefore needed—one that integrates ecological sustainability with human empowerment in discussions of coastal entrepreneurship.

This chapter seeks to fill that gap. Its central focus is on positioning human resource empowerment as the core of sustainable entrepreneurship in coastal regions. The argument is straightforward: technology and regulations matter, but they are insufficient unless communities themselves have the capacity to turn external pressures into opportunities for growth.

Building on this foundation, the chapter pursues four objectives. First, it maps the ecological, socio-economic, and governance challenges and opportunities that shape coastal entrepreneurship. Second, it examines the strategic role of human resources in linking sustainability policies with local practices. Third, it proposes a conceptual model that places entrepreneurship on the foundation of human empowerment. Finally, it offers policy recommendations for building ecosystems that nurture green enterprises aligned with the SDGs.

The contribution of this chapter is twofold. Theoretically, it integrates the perspectives of human capital, the capability approach, and community-based entrepreneurship into a unified framework for coastal sustainability. Practically, it provides insights for policymakers, NGOs, and practitioners to design interventions that are not only economically viable but also ecologically sound and socially inclusive. Methodologically, the analysis draws on a qualitative approach that combines a thematic literature review with empirical insights from Indonesian case studies. The findings aim to demonstrate how human empowerment can translate into concrete outcomes, such as improved catch efficiency, higher household income, ecosystem recovery, and greater gender participation.

The chapter is structured as follows. It begins by mapping the ecological pressures and socio-economic constraints facing coastal communities. It then analyzes the role of human resources in fostering adaptation and innovation. Next, it develops a conceptual model of eco-friendly entrepreneurship. Finally, it concludes with policy recommendations for governments, civil society, and the private sector to promote inclusive and sustainable coastal development.

CHALLENGES AND OPPORTUNITIES FOR ENTREPRENEURSHIP IN COASTAL COMMUNITIES

Coastal entrepreneurship in Indonesia thrives in a paradoxical landscape. For hundreds of years, the ocean, coral reefs, and mangrove forests have been the primary sources of livelihood for coastal communities. However, pressure on these ecosystems is now intensifying. Overfishing has depleted stocks to critical levels, mangrove and coral degradation is widespread, and plastic debris piling up along coastlines exacerbates the damage. Climate change compounds the complexity of the problem, from sea level rise to unpredictable rainfall to increasing sea temperatures. All of these factors reduce biodiversity and make small-scale fishing businesses increasingly vulnerable (FAO, 2022; Hoegh-Guldberg et al., 2019).

Challenges also arise in the social and economic spheres. Limited access to formal finance makes fishermen dependent on high-interest loans from middlemen. Lack of infrastructure and low levels of education hinder innovation and the adoption of new technologies. Distribution chains dominated by intermediaries squeeze small producers into low-value supply chains (Masni et al., 2024). From a regulatory perspective, overlapping policies, weak oversight, and a lack of recognition of local wisdom, such as the marine sasi (*sea sasi*) in Maluku, reduce the effectiveness of resource governance (Bappenas, 2023).

Nevertheless, opportunities continue to emerge. The global market for environmentally friendly fishery products, sustainable aquaculture, and conservation-based tourism is showing an upward trend. The organic shrimp program in Sidoarjo and mangrove ecotourism in Bali demonstrate that conservation can go hand in hand with improving community welfare (Bayuningsih et al., 2024). Meanwhile, digital innovations such as online seafood marketing platforms and virtual training are opening up new market access for fishermen in remote areas while improving their entrepreneurial skills (Areche et al., 2024).

Community-based approaches are increasingly demonstrating their relevance. The resource management council in Maluku, which combines customary rules and formal regulations, demonstrates that participatory governance can strengthen accountability (Fabinyi, 2022). Women's cooperatives in Nusa Tenggara, which process seaweed into processed foods and environmentally friendly cosmetic products, not only increase income but also strengthen women's social roles. All of these experiences confirm that the success of sustainable coastal entrepreneurship requires an integrated strategy that combines ecological conservation, social justice, and inclusive economic development. By transforming challenges into opportunities for innovation, coastal communities can emerge as a vital driver of sustainable blue economy development in Indonesia.

Table 1. Challenges and Opportunities for Coastal Entrepreneurship in the Context of the Blue Economy

Dimensions	Main Challenges	Strategic Opportunities	Indonesian Case Study
Ecological	Overfishing, coral reef degradation, and climate change (FAO, 2022; Hoegh-Guldberg et al., 2019).	Adopting environmentally friendly technology (selective) fishing, eco-aquaculture, and community-based conservation.	East Java: Selective fishing gear training reduces *bycatch* by 20% and increases efficiency (Mochklas et al., 2024).
Socioeconomic	Limited access to formal credit (only 23% of small-scale fishermen receive it); low digital literacy (Bappenas, 2023; Masni et al., 2024).	Green microcredit schemes; use of e-commerce for marketing marine products; business diversification.	Nusa Tenggara: seaweed women's cooperative produces value-added derivative products and increases margins by 25–30% (Areche et al., 2024).
Gender & Inclusion	Women are marginalized from cooperative leadership, have a double workload, and the younger generation participates minimally (Areche et al., 2024; Karataş-Özkan et al., 2024).	Women's empowerment through cooperatives; youth participation in coastal digital innovation.	Nusa Tenggara: Women's cooperatives strengthen household bargaining positions through diversification and collective marketing (Areche et al., 2024).
Institutional	Cooperatives are weak, Minapolitan programs are often unsustainable, and accountability is minimal (Fudge et al., 2023).	Strengthening cooperatives based on transparent rules; multi-actor collaboration (campuses, NGOs, private sector).	Bali: Mangrove ecotourism succeeds due to good cooperative governance and local participation (Bayuningsih et al., 2024).
Technology & Innovation	Limited digital infrastructure; high initial costs of technology adoption.	Coastal business incubation; technology transfer through universities and NGOs; IoT and e-commerce.	Sulawesi & Maluku: a cooperative-based *eco-fishery project* using environmentally friendly technology to reduce destructive practices (Masni et al., 2024).

Source: (FAO, 2022; Hoegh-Guldberg et al., 2019; Mochklas et al., 2024; Areche et al., 2024; Karataş-Özkan et al., 2024; Fudge et al., 2023; Masni et al., 2024; Bayuningsih et al., 2024 et al., 2024; Bappenas, 2023)

Table 1 summarizes how ecological, socioeconomic, gender, institutional, and technological dimensions interact to shape the realities of coastal entrepreneurship.

Each dimension presents a set of persistent barriers but also reveals avenues for innovation when communities, governments, and supporting institutions act strategically.

From an ecological standpoint, the main challenges are overfishing, coral reef degradation, and climate-related pressures (FAO, 2022; Hoegh-Guldberg et al., 2019). These processes undermine marine biodiversity and reduce the productivity of coastal livelihoods. Yet, these same threats have stimulated experiments with environmentally friendly practices such as selective fishing gear, eco-aquaculture, and community-led conservation. The case from East Java illustrates this point: after training programs introduced selective nets, local fishers not only reduced bycatch by nearly 20 percent but also improved their efficiency (Mochklas et al., 2024). Here, ecological vulnerability was transformed into an opportunity for sustainable practice.

The socioeconomic dimension reflects persistent financial exclusion and limited digital literacy. Only a fraction of small-scale fishers in Indonesia have access to formal credit (Bappenas, 2023), forcing many to depend on exploitative middlemen. This restricts their ability to move up the value chain. At the same time, low familiarity with digital tools keeps many households locked into local markets. Yet innovative responses are emerging: green microfinance programs and digital marketing platforms have allowed communities to diversify income and increase profit margins. Women's cooperatives in Nusa Tenggara, for instance, transformed raw seaweed into processed goods that fetch 25–30 percent higher returns (Areche et al., 2024).

The gender and inclusion dimension highlights how women and youth remain underrepresented in leadership and entrepreneurial decision-making. Women often juggle domestic work and economic activity, while younger generations are reluctant to enter fisheries, perceiving them as less rewarding compared to urban employment (Karataş-Özkan et al., 2024). These trends risk excluding significant portions of the population from coastal development. However, inclusive approaches are beginning to reverse this pattern. Women-led cooperatives and youth-driven digital innovation projects in coastal areas have expanded participation and strengthened community bargaining power (Areche et al., 2024).

At the institutional level, many government initiatives have underperformed. Programs such as the Minapolitan policy often face challenges of fragmented coordination, limited accountability, and insufficient community ownership (Fudge et al., 2023). Cooperatives, too, often exist in name only, lacking transparency and operational strength. Yet successful cases from Indonesia and abroad demonstrate alternatives. In Bali, mangrove-based ecotourism has thrived precisely because cooperatives were well managed and inclusive of community participation (Bayuningsih et al., 2024). Similarly, in the Philippines, CBCRM programs that combine technical support with clear governance rules have proven effective in sustaining both livelihoods and ecosystems (Fabinyi, 2022).

Finally, the technology and innovation dimension presents a paradox. High upfront costs and weak infrastructure hinder adoption of modern tools in many coastal areas. Yet where technology transfer has been supported through universities, NGOs, or incubator programs—it has significantly improved sustainability outcomes. Projects in Sulawesi and Maluku, for example, have demonstrated that cooperative-based eco-fisheries using environmentally friendly equipment can reduce destructive practices and strengthen household incomes (Masni et al., 2024).

Taken together, Table 1 makes it clear that coastal entrepreneurship is shaped by interlocking systems. Ecological decline exacerbates economic precarity, financial exclusion reinforces inequality, and weak institutions undermine collective action. At the same time, each constraint holds the potential for transformation when communities are supported with the right mix of knowledge, resources, and institutional backing. The key lesson is that opportunities do not arise in isolation they emerge precisely through efforts to address and overcome challenges in a holistic, integrated way.

THE STRATEGIC ROLE OF HUMAN RESOURCES

In discussions of coastal entrepreneurship, the focus often falls on ecological resources or financial capital. Yet, the decisive factor that determines whether enterprises survive and thrive is the quality and empowerment of human resources. Skills, knowledge, and collective organization function as the critical bridge that links ecological sustainability with economic viability. Without capable and motivated people, investments in technology or infrastructure remain underutilized and institutional reforms falter.

Coastal communities operate in environments marked by uncertainty—rising sea levels, shifting fish stocks, market volatility, and institutional weaknesses. Within such contexts, entrepreneurship cannot be sustained by capital inputs alone. It depends on the ability of individuals and groups to adapt, learn, and innovate. Training in selective fishing, for example, does more than improve efficiency; it reduces bycatch, ensures ecological balance, and stabilizes household income (Mochklas et al., 2024). Similarly, initiatives that encourage women to move beyond processing into leadership positions expand both economic opportunities and social equity (Areche et al., 2024).

Seen from this perspective, human resources are not passive recipients of external aid. They are active agents who shape the direction of entrepreneurial development. Leadership, organizational skills, and the ability to coordinate collective action allow communities to negotiate better prices, build trust with external partners, and embed sustainability principles in everyday practice.

The empowerment of human resources is often equated with *capacity building*, usually measured by the number of training workshops or technical sessions conducted. While such programs are important, they frequently emphasize inputs rather than outcomes. A more strategic view requires shifting attention to tangible results. Economically, this means households should experience higher and more stable earnings. Ecologically, empowerment should manifest in reduced destructive practices and improved ecosystem restoration. Institutionally, the benefits should be visible in stronger cooperatives, inclusive governance, and transparent management practices.

This orientation aligns with Human Capital Theory (Becker, 1993), which stresses the link between skill investment and productivity, as well as Sen's Capability Approach (1999), which underscores the freedom and agency to convert resources into meaningful achievements. What matters is not how many programs are delivered but whether people gain the capacity to transform challenges into sustainable opportunities.

Strong human resources also underpin institutional reform. Many cooperatives collapse because members lack financial literacy, leadership skills, or the confidence to enforce accountability. By contrast, when communities possess these capacities, organizations become resilient and effective. In Bali, for example, ecotourism projects succeeded because cooperatives combined technical knowledge with inclusive decision-making, creating trust and long-term commitment (Bayuningsih et al., 2024). This shows that institutional strength does not emerge automatically; it is built through continuous investment in people.

Human resources, therefore, must be viewed not simply as labor inputs but as the foundation of adaptive and sustainable entrepreneurship. Technical expertise, leadership capacity, and cultural knowledge interact to create innovative solutions suited to local realities. When coastal residents are empowered in this way, ecological protection and economic development cease to be competing priorities they reinforce one another.

In short, the strategic role of human resources lies in their ability to transform policies and technologies into lived practice. They connect global sustainability goals with everyday decisions in coastal households and organizations. As such, investing in human empowerment is not an optional add-on but the central pathway through which the blue economy can become genuinely inclusive and resilient.

Table 2. Evaluation Dimensions of Human Resource Empowerment in Coastal Entrepreneurship

Dimensions	Key Indicators	Practical Example
Economy	Increased revenue, efficiency, and diversification	Selective fishing gear training, e-commerce, and cooperative management
Ecology	Sustainable practices, ecosystem restoration	Ecotourism guide training, mangrove restoration, and reduction of juvenile fishing
Institutional	Strengthening cooperatives, inclusive governance	Leadership rotation, financial literacy, and collective procurement

Source: (Bappenas, 2023; Mochklas et al., 2024; Fudge et al., 2023; Bayuningsih et al., 2024; Areche et al., 2024)

Table 2 shows that the conceptual framework for sustainable coastal entrepreneurship is built on three interrelated pillars: ecological sustainability, socio-economic progress, and human resource development. The ecological pillar serves as the primary foundation, emphasizing biodiversity conservation, preventing overexploitation, and rehabilitating damaged ecosystems. Without these safeguards, critical sectors such as fisheries, aquaculture, and tourism will face long-term vulnerability (FAO, 2022).

The socio-economic dimension plays a role in creating a fair and inclusive business climate. Transparent market access, stable household incomes, and institutional support through cooperatives or resource management councils can strengthen the bargaining position of communities. These efforts also align with national policies such as Minapolitan, which promote economic integration and local empowerment (Bappenas, 2023).

The third pillar, human resources, serves as a bridge between ecological needs and socio-economic aspirations. Technical training, digital literacy, and leadership development enable communities to translate policies into practical practice. Real-life examples include selective fishing gear training in East Java, mangrove ecotourism development in Bali, and women's seaweed cooperatives in Nusa Tenggara (Mochklas et al., 2024; Bayuningsih et al., 2024).

This framework emphasizes sustainability as a cyclical process. A healthy ecology creates economic opportunities, empowered local actors drive innovation, and business profits are reinvested in conservation and capacity building. In this way, coastal entrepreneurship is oriented not only toward economic growth but also toward strengthening social justice and environmental resilience.

Human Resource Empowerment Case Study

Case studies in Indonesia demonstrate that strengthening human resource capacity can transform the concept of sustainability from mere jargon into concrete practices embedded in daily life. In this context, empowerment is not simply understood as a process of skills transfer. It encompasses building self-confidence, developing leadership skills, and the ability to innovate amidst uncertain times. This process enables coastal communities not only to survive but also to reimagine the relationship between entrepreneurship and ecological responsibility (Fudge et al., 2023).

On the coast of East Java, for example, fishermen in a cooperative have begun using selective fishing technology. This change initially reduced bycatch and increased efficiency. However, the benefits extend beyond the ecological aspect. The training program also incorporates financial literacy and digital marketing. With these skills, fishermen can access more transparent price information, negotiate collectively, and reduce dependence on middlemen. This combination of strategies strengthens household income stability, even when marine commodity prices fluctuate on the global market (Mochklas et al., 2024). This case highlights that ecological success will only be sustainable if coupled with managerial and digital capacity.

A different approach is evident in Bali, where local communities are developing ecotourism based on mangrove ecosystems. Residents are not only equipped with skills as tour guides but also taught how to convey environmental messages using local cultural narratives. Digital promotion training enables them to reach tourists beyond the surrounding area. The economic results are tangible, with increased household incomes. More importantly, a portion of this income is reallocated to mangrove rehabilitation and community-based waste management (Bayuningsih et al., 2024). Here, improving well-being and environmental conservation are not mutually exclusive, but rather mutually reinforcing.

Further east, in Nusa Tenggara, a women-led seaweed cooperative demonstrates how empowerment can transform both markets and social structures. With external support, cooperative members shifted from selling raw seaweed to producing eco-labeled snacks and cosmetics. Profits increased, while women's roles as decision-makers within the cooperative and within the community became more visible. This example demonstrates that sustainability strategies grounded in gender inclusion not only enhance household economic resilience but also address long-standing inequalities (Areche et al., 2024).

All these cases, combined, point to one crucial understanding: sustainable coastal entrepreneurship relies heavily on comprehensive human resource development. Technical knowledge, business acumen, and digital literacy are crucial, but they need to be supported by institutions that ensure accountability and participation. Without such a foundation, green business models remain mere talk. Conversely, when implemented with solid institutional support, sustainability can be realized as a concrete practice that strengthens ecosystems and enhances the socio-economic resilience of communities (Fabinyi, 2022; Bappenas, 2023).

Table 3. Case Study of Human Resource Empowerment in Coastal Communities

Location	HR Intervention	Key Results
Nusa Tenggara	Technical assistance for women's cooperatives	Margins 20–30% higher, revenues increased, and women's participation increased.
Bali	Ecotourism guide training, product design	Household incomes increase by 25–30%, and mangrove restoration funds are available.
General	Digital training, e-commerce	Stable prices, reduced dependence on middlemen, wider market access

Source: (Mochklas et al., 2024; Areche et al., 2024; Bayuningsih et al., 2024; Masni et al., 2024)

Table 3 demonstrates the integrated policy direction designed to strengthen sustainable entrepreneurship in coastal areas. This approach does not view ecology, socio-economics, and institutions as stand-alone dimensions, but rather as parts of an interconnected and mutually reinforcing system (Bappenas, 2023).

Ecologically, stricter monitoring of natural resource utilization and the restoration of degraded ecosystems are crucial. Mangrove rehabilitation programs, marine habitat protection, and incentives for environmentally friendly practices are positioned as long-term investments. These efforts not only maintain the sustainability of fisheries and aquaculture but also ensure the attractiveness of coastal tourism (FAO, 2022; Bayuningsih et al., 2024).

Socio-economically, the focus is on strengthening human capacity. Field experience shows that technical training, digital literacy, and cooperative management, when combined with access to microfinance and market networks, can increase household incomes while strengthening community resilience. This initiative also opens up broader participation for women and young people to take leadership roles, as seen in the seaweed cooperatives in Nusa Tenggara, which have successfully expanded their markets to include certified derivative products (Mochklas et al., 2024; Areche et al., 2024).

Institutional aspects, emphasis is placed on transparent and participatory governance. Co-management schemes that combine local wisdom, such as the practice of sasi laut in Maluku, with formal regulations, have been shown to strengthen accountability and ownership in resource management. Collaboration with universities, NGOs, and the private sector also encourages knowledge transfer and program sustainability beyond donor funding cycles (Fabinyi, 2022).

This strategic integration shifts the measurement of coastal development success from mere economic output to broader indicators: ecological resilience, social equity, and long-term sustainability. When environmental protection, community empowerment, and strong institutions work hand in hand, coastal entrepreneurship not only survives economically but also thrives as an inclusive and environmentally friendly practice.

ENVIRONMENTALLY FRIENDLY ENTREPRENEURSHIP MODEL

Designing a sustainable model of entrepreneurship in coastal regions requires more than cosmetic "green" adjustments to conventional business practices. What is needed is an integrated strategy that links ecological stewardship, human resource empowerment, and institutional reinforcement into a single framework. The ultimate objective is not only to generate income but also to ensure that communities can prosper without undermining the ecosystems that sustain them. Such a model should remain flexible, capable of being tailored to the cultural traditions, ecological conditions, and institutional capacities of each locality (Bappenas, 2023).

At the foundation of this approach lies ecological integrity. No entrepreneurial activity can truly be sustainable if marine habitats are collapsing due to destructive fishing, coral bleaching, or unchecked pollution. Conservation practices such as mangrove rehabilitation in Demak, selective fishing gear introduced in East Java, or community-led waste reduction in Lombok are not merely environmental projects—they are investments that secure both biodiversity and economic continuity (FAO, 2022; Mochklas et al., 2024). These initiatives show that protecting ecosystems strengthens the resource base for long-term business viability.

The second pillar is the empowerment of local communities. Raising awareness alone is insufficient unless people also acquire the skills, confidence, and autonomy to act. Programs that provide training in digital marketing, financial literacy, or cooperative management allow fishers, processors, and traders to innovate and adapt. In Nusa Tenggara, for instance, women's cooperatives that shifted from raw seaweed sales to processed snacks and cosmetics not only improved profits but also enhanced women's leadership roles in their villages (Areche et al., 2024). Empowerment in this sense is about enabling agency communities creating their own solutions rather than depending indefinitely on external interventions.

Equally critical is institutional support. Even capable entrepreneurs require an enabling environment to succeed. Reliable policies, fair market access, and transparent governance are essential in translating ecological and human resource investments into durable outcomes. Collaborative examples, such as co-managed fisheries in Maluku that combine state regulation with *sasi laut* customs, illustrate how local wisdom can reinforce formal systems and generate accountability (Fabinyi, 2022). Partnerships with universities and NGOs further expand access to research, innovation, and funding, ensuring that small-scale actors are not left behind.

What makes this model distinct is its adaptive and cyclical character. Instead of treating sustainability as a final target, it recognizes continuous feedback loops. Revenues from ecotourism in Bali, for example, have been reinvested into mangrove rehabilitation, which in turn enhances the tourism product itself (Bayuningsih et al.,

2024). In this virtuous cycle, ecological restoration and business growth reinforce one another, creating resilience against both environmental shocks and market volatility.

Ultimately, sustainable coastal entrepreneurship arises not from choosing between economic gain or environmental care, but from weaving ecological responsibility, empowered communities, and strong institutions into a single strategy. When these components operate in concert, coastal societies are better equipped to withstand disruptions while actively shaping an inclusive and enduring blue economy.

Conceptual Foundation: Green Entrepreneurship in a Coastal Context

In various coastal areas of Indonesia, the idea of green entrepreneurship has grown from the realization that economic prosperity and environmental sustainability do not have to be positioned as mutually exclusive. Instead, they can coexist and support each other. For fishing families, fish farmers, and tourism operators, business success is measured not only by financial returns but also by the ability to maintain the marine ecosystems that support daily life. Thus, green entrepreneurship is not merely a theoretical discourse, but a real need that determines the sustainability of coastal communities (FAO, 2022; Bappenas, 2023).

This foundational thinking is closely linked to two important academic traditions. First, human capital theory emphasizes that education, health, and skills are the primary drivers of increased productivity and innovation (Becker, 1993). Second, the capability approach highlights the importance of individual freedom to transform resources into meaningful achievements (Sen, 1999). In the local context, sustainability can only be achieved if fishermen are skilled at using environmentally friendly fishing gear, women are able to master digital marketing strategies, or the younger generation has the financial literacy to manage risk (Mochklas et al., 2024).

This thinking is then reinforced through the implementation of green business models and circular economy principles. Both promote efficiency, resource reuse, and waste reduction. Real-world practices can be found in community-based recycling initiatives in Surabaya, sustainable shrimp farming programs in East Java, and the development of seaweed derivative products in Nusa Tenggara involving women's groups (Bayuningsih et al., 2024). All of these examples demonstrate that sustainability is not an additional burden, but rather a strategy that provides long-term benefits.

The community-based entrepreneurial dimension should also not be overlooked. The sea sasi tradition in Maluku, for example, demonstrates how social norms and collective trust can create effective resource management mechanisms (Fabinyi, 2022). Field experience demonstrates that when global theory is combined with local values, green entrepreneurship transforms into practical practice. It not only maintains ecological sustainability but also strengthens social solidarity and expands the economic resilience of Indonesian coastal communities.

Principles of Environmentally Friendly Entrepreneurship

Efforts to develop sustainable businesses in coastal zones are not only about profitability but also about aligning economic practices with ecological stewardship and community resilience. The guiding principles are intended to be practical rather than theoretical, offering pathways for local actors to manage marine resources responsibly while securing their livelihoods (Bappenas, 2023).

The first and perhaps most fundamental principle concerns ecological responsibility. Coastal ecosystems—mangroves, seagrass beds, and coral reefs—are highly sensitive and limited in their capacity to regenerate. Entrepreneurs who invest in habitat restoration, adopt selective fishing gear, or establish waste collection systems are simultaneously protecting the environment and ensuring the durability of their enterprises. Without healthy ecosystems, long-term profitability is unsustainable, a reality already visible in declining fisheries across parts of Sulawesi and North Sumatra (FAO, 2022).

A second principle emphasizes human capacity building. Access to natural resources alone does not guarantee resilience; communities must also acquire skills to adapt and innovate. Training in financial literacy, digital marketing, or aquaculture management enables households to transform risks such as climate variability into new opportunities. In East Java, for instance, cooperatives that received combined training in technical fishing methods and digital promotion managed to increase catch efficiency while accessing broader markets (Mochklas et al., 2024).

The third principle relates to social inclusion. Women, youth, and marginalized groups are often at the heart of processing, trading, and household financial management, yet their voices in leadership remain limited. Experiences from Nusa Tenggara show that when women's cooperatives were supported to diversify seaweed into snacks and cosmetics, not only did incomes rise but community decision-making became more equitable (Areche et al., 2024). Inclusivity, therefore, is not simply a matter of justice but also a driver of resilience and trust in collective resource management.

A fourth principle highlights adaptive innovation. Sustainability is not achieved through rigid templates but through the capacity to combine traditional knowledge with contemporary tools. Examples include integrating aquaculture with mangrove rehabilitation in Central Java or using mobile applications to link small-scale fishers directly with ethical seafood buyers in Jakarta's urban markets (Bayuningsih et al., 2024). These hybrid approaches show how cultural traditions and modern technologies can complement one another.

Finally, sustainable entrepreneurship depends on institutional support and accountability. Strong governance, clear regulations, and partnerships with universities, NGOs, or private actors create the enabling environment for eco-enterprises to flourish. Accountability mechanisms, such as community monitoring groups in Maluku, help ensure commitments to conservation are honored while granting local people a say in shaping outcomes (Fabinyi, 2022). These principles form a holistic framework. By combining ecological care, skill development, social inclusion, innovation, and institutional backing, coastal communities are better positioned to achieve economic stability while safeguarding the natural and social systems that will sustain future generations.

Empirical Evidence in Indonesia

Case studies in Indonesia demonstrate that sustainability only takes on real meaning when translated into practical strategies implemented by communities themselves. Investments in human resources—whether through skills development, increased self-confidence, or institutional support—transform abstract environmental ideas into entrepreneurial practices that can provide livelihoods while preserving ecosystems. The long-term sustainability of these efforts depends on the synergy between technical capacity, managerial skills, and accountable governance. These three aspects reinforce each other in a sustainable and mutually beneficial cycle (Bappenas, 2023; Fudge et al., 2023).

In East Java, for example, a training program aimed at small-scale fishing cooperatives introduced the use of selective fishing gear. As a result, bycatch was reduced by almost 20 percent and production efficiency increased. However, this intervention went beyond ecological aspects. Fishermen also received training in financial literacy and digital marketing. This knowledge enabled them to negotiate directly with buyers, eliminating their complete dependence on middlemen. In this way, family incomes become more stable despite the frequent fluctuations in global marine commodity prices (Mochklas et al., 2024). This case demonstrates that ecological awareness must be combined with business acumen and digital skills for long-term sustainability.

A different approach is evident in Bali, where local communities transformed degraded mangrove areas into ecotourism destinations. Residents received training not only as tour guides but also in environmental interpretation, visitor management, and online promotion. Consequently, household incomes increased by around 25–30 percent. More importantly, a portion of the profits was allocated to mangrove rehabilitation and community-based waste management. This case demonstrates how entrepreneurship can create a virtuous cycle, where economic success strengthens conservation, and conservation, in turn, maintains business sustainability (Bayuningsih et al., 2024).

Another example comes from Nusa Tenggara, where a women-led seaweed cooperative offers a new perspective on empowerment. With support in production technology and cooperative governance, members were able to transition from selling raw materials to selling value-added processed products, such as snacks and eco-certified cosmetics. This diversification increases profits while strengthening women's positions in leadership and decision-making at the community level. This case demonstrates that empowerment is not only related to economic growth but also to changing gender relations and strengthening community social resilience (Areche et al., 2024).

This case study emphasizes that sustainable coastal enterprise development will only succeed if ecological responsibility, human empowerment, and institutional support go hand in hand. Technical knowledge alone is not enough; it must be combined with financial literacy, digital skills, and inclusive leadership. The integration of these dimensions enables coastal enterprises to generate stable incomes, strengthen social equity, and maintain the long-term sustainability of marine ecosystems (Fabinyi, 2022; Masni et al., 2024).

Table 4. Case Study of Human Resource Empowerment and Environmentally Friendly Entrepreneurship in Coastal Communities

Location	HR Intervention	Key Results
East Java	Selective fishing gear training, digital marketing	Bycatch reduced by 20%, CPUE increased, price stable
Bali	Ecotourism guide training, interpretation, safety, and digital promotion	Household income increases by 25–30%, mangrove restoration funding, and plastic waste reduction
Nusa Tenggara	Technical assistance for seaweed processing	Diversification of environmentally friendly products, increasing margins, and increasing female leadership

Source: (Mochklas et al., 2024; Bayuningsih et al., 2024; Areche et al., 2024)

Table 4 shows that environmentally friendly entrepreneurial practices can only thrive if human resource development goes hand in hand with economic innovation and ecological awareness. In East Java, for example, fishermen's cooperatives introduced to selective fishing technology successfully reduced bycatch while increasing productivity. However, the greatest benefits emerged when ecological interventions were combined with financial literacy training and digital marketing. Fishermen were able to negotiate prices directly, reducing dependence on middlemen, and maintaining income stability despite fluctuating market prices. A similar trend was seen in Bali, where a mangrove ecotourism program not only increased household income by 25–30 percent but also encouraged reinvestment in ecosystem rehabilitation and community-based waste management. Both cases emphasize that sustainability stems from the integration of ecological innovation, managerial skills, and digital literacy, not from either element alone (Mochklas et al., 2024; Bayuningsih et al., 2024).

These entrepreneurial practices demonstrate the importance of institutional support capable of maintaining program accountability and sustainability. Without a clear policy framework, technical and economic achievements are vulnerable to stalling. Conversely, with strong institutions, as demonstrated by community-based initiatives in Bali and cooperatives in East Java, economic benefits can go hand in hand with environmental improvements. This pattern creates a "double dividend": community well-being improves, while fragile coastal ecosystems receive sustainable protection. Therefore, green entrepreneurship in coastal areas is not only an instrument for profit generation but also a foundation for inclusive development that integrates social, economic, and ecological aspects (Bappenas, 2023; Masni et al., 2024)..

In Nusa Tenggara, women-led cooperatives in seaweed processing underscore the power of empowerment and inclusivity. With targeted support in both production techniques and cooperative governance, members expanded beyond raw exports into value-added foods and eco-certified cosmetics. The result was not only higher profitability but also an increase in women's participation in leadership and community decision-making. This case demonstrates that sustainable business models can simultaneously advance ecological goals, strengthen economic performance, and correct gender imbalances (Areche et al., 2024).

Taken together, these cases affirm a key principle: the impact of technical, managerial, and digital skills is greatest when embedded within institutions that enforce transparency and ecological responsibility. Without strong institutional backing, sustainability risks remaining rhetorical. But when such integration occurs, communities are better able to achieve ecological resilience, economic security, and inclusive growth (Fabinyi, 2022).

POLICY AND RESEARCH IMPLICATIONS

The findings carry important implications for policy. An entrepreneurship model rooted in human resource development requires that human empowerment, governance, and ecological care be treated as interdependent. National programs—such as *Minapolitan*—should not be judged solely on economic outcomes. Broader indicators are needed, including household income stability, improvements in CPUE, women's participation in cooperatives, and tangible results in ecosystem restoration. By widening the scope of evaluation, policies will better capture the multidimensional character of sustainability (Bappenas, 2023).

Pilot experiences in East Java, Bali, and Nusa Tenggara suggest that replication is possible, but scaling up demands continuous investment. Strengthening cooperative governance, expanding digital infrastructure, and embedding training systems at the local level remain priorities. With these supports, eco-entrepreneurship can move from isolated cases to a nationwide strategy that balances prosperity with ecological integrity and social equity. By broadening the criteria for success, policies can better capture the multidimensional nature of sustainability and ensure that economic growth is accompanied by social empowerment and ecological resilience. This will enable policymakers to understand the extent to which human resource interventions truly improve community capacity while maintaining the balance of coastal ecosystems. Furthermore, pilot experiences in East Java, Bali, and Nusa Tenggara demonstrate potential for replication. However, ongoing investment in training systems, strengthening cooperative governance, and digital infrastructure is needed for the program to scale effectively.

From a research perspective, future agendas should explore the long-term impacts of human resource interventions, including the gender dimension of eco-entrepreneurship, and how digital ecosystems can accelerate the development of environmentally friendly businesses. The emphasis on human resources at the core of this model emphasizes that green entrepreneurship is not simply a response to market pressures, but rather a systemic driver of inclusive sustainability. With adequate human resource capacity, coastal communities can transform ecological, social, and institutional challenges into innovative opportunities, broaden inclusion, and strengthen collective resilience.

Empirical evidence in Indonesia shows that human resource empowerment transforms sustainability principles into tangible, measurable outcomes, both economically, socially, and ecologically. The success of this model depends on a combination of skilled and empowered human resources, equitable institutional support, and consistent practices of ecological reciprocity. This approach directly supports the achievement of SDGs 12, 13, and 14, while positioning coastal communities as active actors in the global blue economy. The following section will explore how human resource empowerment impacts entrepreneurial performance and sustainable development outcomes, linking micro-initiatives to macro-policies, and affirming the position of coastal communities as drivers of blue economy transformation.

Policy Recommendation

Strengthening *ecopreneurship* in Indonesia's coastal areas requires policies based on a holistic perspective, not just partial interventions. Research shows that community practices are often treated as separate entities. These three aspects should be woven into an integrated framework that addresses ecological vulnerability, socio-economic inequality, gender marginalization, weak institutions, and technological limitations. Such an integrative approach not only closes policy gaps but also opens up opportunities for sustainable growth through the ecotourism sector, the use of digital platforms, and environmentally friendly resource management. If managed well, these opportunities can support inclusive development while protecting increasingly vulnerable marine ecosystems. Several policy recommendations that need to be implemented are:

Financial Incentives and Support

In many coastal areas in Indonesia, the main obstacle to sustainable business development is not only limited capital, but also the suitability of financing instruments to local realities. The government, along with financial institutions, cannot simply rely on conventional credit schemes, which are often rigid and difficult for small-scale fishermen and micro-entrepreneurs to access. Therefore, breakthroughs are needed in the form of more flexible alternative financing instruments, such as green microfinance, blended finance, and technology-based grants. With a more contextual approach, financial access will not only encourage innovation but also strengthen the competitiveness of coastal businesses while maintaining the sustainability of marine ecosystems.

Strengthening Human Resource Capacity

Coastal entrepreneurship development is not simply about capital; it must also place people at the heart of the development process. Furthermore, field experience shows that without capacity building, many coastal economic initiatives struggle to sustain themselves in the long term. Therefore, public policy needs to be directed toward a more holistic approach, one that combines local wisdom with contemporary training strategies, proven to strengthen community capacity while maintaining business sustainability.

Blue Economy Integration in Regional Planning

Institutional fragmentation often hinders cross-sector coordination. Policies should encourage integration between the fisheries, tourism, and MSME sectors within regional planning frameworks, such as the Regional Medium-Term Development Plan, by establishing performance indicators aligned with the SDGs. Such integration fosters operational synergy, balances economic growth with ecological conservation, and provides a foundation for objective, results-based evaluation.

Development of Local Innovation Ecosystem

Local governments have a crucial role to play in fostering local innovation, particularly in coastal areas vulnerable to environmental and market pressures. This can be achieved by strengthening business incubators, building strong collaborations between universities and communities, and accelerating the transfer of technology relevant to community needs. However, field experience demonstrates that innovation should not be solely focused on achieving short-term economic gains. Instead, strengthening the capacity of digital entrepreneurs, processing environmentally friendly products, and implementing appropriate technology must be prioritized to increase competitiveness without sacrificing the sustainability of marine ecosystems.

Strengthening Regulation and Monitoring

Encouraging the growth of sustainable coastal entrepreneurship requires more than just creating economic opportunities; it also requires robust governance. The government needs to enforce clear environmental standards, strengthen community participation mechanisms, and establish a consistent monitoring system. In practice, designed policies must ensure transparency, prevent domination by local elites, and provide an integrated evaluation framework capable of simultaneously assessing economic, social, and ecological impacts.

Implementation Recommendation

Promoting green entrepreneurship in coastal communities requires a phased, adaptive, and collaborative approach that integrates technical, institutional, social, and economic dimensions. In the short term (1–3 years), policies focus on building core capacities through training, access to green microcredit, and local environmental standards. The medium term (4–6 years) emphasizes blue economy integration, cooperative governance, and coastal business incubators, while the long term (7–10 years) prioritizes blended financing, national innovation networks, and adaptive climate regulations to ensure sustainability and resilience (Fudge et al., 2023).

This policy approach highlights the importance of flexibility, recognizing that interventions must evolve in step with local ecological conditions, the skills of human resources, and shifting socio-economic realities. No single actor can achieve this alone. National agencies contribute by setting regulatory frameworks and allocating funds, while local governments are better positioned to implement and monitor initiatives on the ground. Universities and research centers play a complementary role through training and technology transfer. Civil society groups and NGOs foster participation and ensure that social inclusion is not overlooked. At the same time, private companies can strengthen the process through corporate social responsibility programs, sustainable supply chain practices, and the diffusion of new technologies (Areche et al., 2024). When these actors work in concert, they reinforce one another—improving skills, institutional performance, financial access, and innovation—while aligning with multiple Sustainable Development Goals (SDGs), such as human resource development (SDGs 8, 14), financial inclusion (SDGs 5, 8, 12), blue economy planning (SDGs 8, 13, 14), innovation ecosystems (SDGs 9, 12, 17), and participatory governance (SDGs 12, 14, 17) (Fabinyi, 2022). Such integration enables coastal communities to turn persistent local challenges into opportunities for inclusive and sustainable development.

Future Policy Directions

Looking forward, coastal policy frameworks need to embrace a holistic vision of sustainability—one that is simultaneously economic, social, and ecological, while also adaptive to rapidly changing local and global contexts. Green entrepreneurship policies should be grounded in empirical experience, informed by inclusive development theory, and responsive to environmental, demographic, and market transitions. Such strategies aim to ensure that economic development benefits local communities while preserving the ecosystems that sustain their livelihoods.

Addressing climate change becomes a key component of this approach. Planning for coastal entrepreneurship must incorporate actions that enhance resilience against rising sea levels, extreme weather events, and fluctuations in marine resources, while promoting sustainable practices. Equally vital is the adoption of digital tools and innovative technologies. Online platforms can broaden market reach, improve supply chain monitoring, and increase the value of locally produced goods.

Successfully implementing these strategies requires collaboration across multiple sectors, including government agencies, universities, private enterprises, and community organizations. This kind of partnership not only facilitates faster technology adoption but also magnifies social and ecological benefits. In this framework, eco-entrepreneurship transcends mere economic objectives. By combining environmental stewardship with innovation and inclusive participation, it ensures both the sustainability of natural resources and the well-being of coastal populations, transforming them into active participants in building a resilient and enduring blue economy for the future.

CONCLUSION

This chapter emphasizes the pivotal role of human resources (HR) in promoting environmentally responsible entrepreneurship within Indonesia's coastal communities. The discussion illustrates that the effectiveness of the blue economy depends not only on technology, financial capital, or regulatory frameworks, but fundamentally on the expertise, practical skills, mindset, and organizational capabilities of individuals and community groups working together.. Empowered HR can transform ecological, social, and institutional challenges into innovative, inclusive, and sustainable opportunities. Based on these findings, several key conclusions can be formulated as follows:

a) Human resources' strategic role lies in their ability to bridge sustainability goals with concrete results in the economic, ecological, and institutional spheres. Improving technical, digital, and managerial capacity not only strengthens productivity but also supports sustainable resource management and strengthens local institutions that underpin the survival of coastal communities.
b) The eco-friendly entrepreneurship model in coastal communities centers on three essential pillars: developing human resource competencies, strengthening institutional ties, and implementing the principle of ecological reciprocity. Field experience shows that economic and ecological sustainability are not separate goals; both can be achieved simultaneously if human resources are empowered comprehensively and supported by an institutional structure responsive and adaptive to local dynamics.
c) The thematic policy recommendations emphasize five complementary strategic areas: strengthening human resource capacity, providing financial incentives, integrating the blue economy into regional planning, developing local innovation ecosystems, and strengthening regulations and monitoring mechanisms. The phased implementation approach, involving various actors, not only strengthens cross-sector collaboration but also ensures active community participation, making the policies more adaptive, inclusive, and impactful in the long term.
d) The human resource-based approach significantly contributes to achieving several Sustainable Development Goals (SDGs), especially SDGs 8, 12, 13, 14, and 17. This strategy creates environmentally friendly jobs, encourages sustainable production practices, protects marine ecosystems, and strengthens cross-sector partnerships supporting inclusive and sustainable development.
e) Book This chapter highlights the importance of integrating human capital theory and the capability approach. Approach, and the blue economy framework with real-world practices. These findings confirm that human resource empowerment

is not merely a normative goal but a key driver of systemic transformation that enables the blue economy to develop inclusively, adaptively, and sustainably.

Overall, coastal communities are now emerging as beneficiaries and central actors driving transformation in the global blue economy. The success of eco-friendly entrepreneurship depends heavily on human resource empowerment, institutional strengthening, financial access, and the creation of an integrated innovation ecosystem. Therefore, human resource capacity development must be a top priority in formulating coastal development policies that are sustainable, responsive to change, and inclusive for all community members.

REFERENCES

Areche, F. O., Dominguez, J. A. J., Huaman, J. T., Yapias, R. J. M., Rivera, T. J. C., Perales, L. D. M., Quispe, J. D. D. H. T., Zea, C. Y. H., Galarza, C. R. C., Rodríguez, A. R., & Areche-Mansilla, D. Y. (2024). Maximizing the Potential of Marine Resources: A Sustainable Approach to High-Value Product Development from Seafood by-Products and Waste. *Current Research in Nutrition and Food Science, 12*(3), 1074–1092. DOI: 10.12944/CRNFSJ.12.3.7

Bappenas. (2023). *Laporan Pembangunan Kelautan Indonesia*. Jakarta: Bappenas.

Bayuningsih, D., Dzulkifli, M., & Muksin, M. (2024). Innovative Approaches in Tourism Development to Revitalize Global Sustainability. *International Journal Of Studies In Social Sciences And Humanities, 1*(1), 1–17.

Becker, G. S. (1993). *Human capital: A theoretical and empirical analysis, with special reference to education* (3rd ed.). University of Chicago Press., DOI: 10.7208/chicago/9780226041223.001.0001

Boons, F., & Lüdeke-Freund, F. (2013). Business models for sustainable innovation: State of the art and steps towards a research agenda. *Journal of Cleaner Production, 45*, 9–19. DOI: 10.1016/j.jclepro.2012.07.007

FAO. (2022). *The State of World Fisheries and Aquaculture 2022*. FAO.

Food and Agriculture Organization. (2022). The State of World Fisheries and Aquaculture 2022. Towards Blue Transformation. In *FAO*. F.A.O., DOI: 10.4060/cc0461en

Fudge, M., Ogier, E., & Alexander, K. A. (2023). Marine and coastal places: Well-being in a blue economy. *Environmental Science & Policy, 144*, 64–73. DOI: 10.1016/j.envsci.2023.03.002

Geissdoerfer, M., Savaget, P., Bocken, N. M. P., & Hultink, E. J. (2017). The circular economy – A new sustainability paradigm? *Journal of Cleaner Production, 143*, 757–768. DOI: 10.1016/j.jclepro.2016.12.048

Hoegh-Guldberg, O., Pendleton, L., & Kaup, A. (2019). People and the changing nature of coral reefs. *Regional Studies in Marine Science, 30*, 100699. DOI: 10.1016/j.rsma.2019.100699

Issifu, I., Alava, J. J., Lam, V. W. Y., & Sumaila, U. R. (2022). Impact of Ocean Warming, Overfishing, and Mercury on European Fisheries: A Risk Assessment and Policy Solution Framework. *Frontiers in Marine Science, 8*(770805), 1–13. DOI: 10.3389/fmars.2021.770805

Karatas-Ozkan, M., Tunalioglu, R., Ibrahim, S., Ozeren, E., Grinevich, V., & Kimaro, J. (2024). Actioning sustainability through tourism entrepreneurship: Women entrepreneurs as change agents navigating through the field of stakeholders. *Central European Management Journal*, *32*(1), 31–56. DOI: 10.1108/CEMJ-12-2021-0159

Masni, H., Hutabarat, Z. S., Rahmadani, C., Yulita, K., Putri, R. A., Nurwati, N., Pianti, U. O., Safitri, R., Ramadani, N., Deskiana, I., Alfarizi, R. P., Putri, R. M., & Yuliana, E. (2024). Strategy for Implementing the Blue Economy Concept for Community Empowerment in Achieving Sustainable Development Goals. *International Journal of Multidisciplinary Sciences and Arts*, *3*(3), 37–43. DOI: 10.47709/ijmdsa.v3i3.3946

Mochklas, M., Maharani, R., Maretasari, R., Panggayudi, D. S., Oktaviani, M., & Muttaqin, R. (2024). Contribution Of Human Resources To Environmentally Friendly Entrepreneurial Models In Coastal Communities To Achieve Sustainable Development Goals (SDGs). *Journal of Lifestyle and SDGs Review*, *4*(2), 1–36. https://doi.org/https://doi.org/10.47172/2965-730X. DOI: 10.47172/2965-730X.SDGsReview.v4.n02.pe01770

Mochklas, M., Panggayudi, D. S., Hafidulloh, H., Iradawaty, S. N., & Pinaraswati, S. O. (2024). The Work Ethos of Kenjeran Beach Traders: A Study of Hard Work, Smart Work, and Sincere Work in the Coastal Tourism Industry. *Journal of Economics, Education, and Entrepreneurship*, *12*(1), 1–23. DOI: 10.26740/jepk.v12n1.p57-78

Sen, A. (1999). *Capabilities and commodities*. Oxford University Press.

United Nations Environment Programme. (2020). *Coral reefs*. Unep.Org. https://www.unep.org/topics/ocean-seas-and-coasts/blue-ecosystems/coral-reefs

World Economic Forum. (2020). *Radically Reducing Plastic Pollution in Indonesia: Multi-Stakeholder Action Plan*. https://www.systemiq.earth/wp-content/uploads/2020/05/NPAP_Indonesia_Action_Roadmap_BahasaLow-1.pdf

KEY TERMS AND DEFINITIONS

Blue Economy: The sustainable use of ocean and coastal resources to promote economic growth, improve community livelihoods, and maintain ecological balance. It includes traditional sectors like fisheries and tourism, as well as emerging areas such as renewable ocean energy and marine biotechnology.

Circular Economy: An economic model focused on reducing waste and maximizing resource efficiency through recycling, reusing, and closed-loop production, particularly relevant in coastal and marine enterprises.

Community-Based Entrepreneurship: A locally driven entrepreneurial approach that emphasizes cooperative structures, inclusive governance, and the integration of socio-cultural values into sustainable business practices.

Eco-Entrepreneurship: Entrepreneurial activities that integrate environmental sustainability into business practices, balancing profit-making with ecological stewardship and social inclusion.

Human Resource Empowerment: A process of enhancing the knowledge, skills, and leadership capacity of individuals and communities to transform environmental, social, and institutional challenges into sustainable entrepreneurial opportunities.

Chapter 6
Marine Tourism:
Niche Segments, Ecological Challenges, and Community Impacts

Melik Onur Güzel
https://orcid.org/0000-0002-6540-0131
Independent Researcher, Turkey

Eşref Ay
https://orcid.org/0000-0003-4092-6425
Independent Researcher, Turkey

ABSTRACT

Marine tourism, one of the fastest-growing segments of global tourism, benefits from natural, cultural, and economic resources, especially in coastal areas. It includes diving, water sports, yacht and cruise travels, and whale watching (Hall, 2001), attracting diverse tourist profiles and supporting destination promotion. However, besides economic gains, it also brings sociocultural and environmental impacts. Mass tourism may lead to ecosystem degradation, sea pollution, and loss of biodiversity (UNEP, 2009). It can also affect the local community's economy, culture, and traditions in both positive and negative ways. This study aims to contribute to the literature by examining these ecological and social effects. The first part discusses the concept, development, and importance of marine tourism, while the following sections evaluate its environmental and social impacts.

DOI: 10.4018/979-8-3373-1777-9.ch006

INTRODUCTION

Marine tourism is one of the fastest growing areas in tourism, which is one of the largest industries in the world, with its natural, cultural and economic resources. This type of tourism, which takes place intensively in coastal areas, includes activities such as diving tourism and water sports, yacht and cruise travels, whale watching, etc. (Hall, 2001). With this rich diversity, marine tourism becomes attractive and interesting for people with different characteristics. It also plays an active role in the promotion of destinations in national and international markets.

However, alongside its economic benefits, marine tourism generates ecological and social impacts that cannot be overlooked. Uncontrolled mass tourism may accelerate ecosystem degradation, biodiversity loss, and marine pollution, while also affecting the livelihoods, culture, and traditions of local communities (UNEP, 2009). This chapter aims to critically examine the social and ecological impacts of marine tourism, synthesizing findings from the literature while highlighting governance, policy, and technological innovations for sustainable management.

LITERATURE REVIEW

Methodological Note

We consulted Web of Science, Scopus, ScienceDirect and Google Scholar (2000–August 2025), prioritising peer-reviewed studies from 2020 onwards. The keywords covered were 'cruise', 'marina', 'diving', 'Blue Flag', 'green/eco ports' (PERS/Green Marine/GPAS), 'shore power', 'carbon-neutral cruise', 'leakage' and 'governance'.

Theoretical Context

Marine tourism studies often rely on tourism impact theories. Butler's Tourism Area Life Cycle (TALC) model (Butler, 2006), Doxey's Irridex model (Doxey, 1975), the carrying capacity approach (Middleton & Hawkins 1998) and Blue Economy Governance (Picken, 2025) provide frameworks to understand destination development, resident responses, and sustainability challenges.

Definition and Scope of Marine Tourism

Marine tourism refers to a specific form of tourism and is the tourist's search for a more direct and more permanent contact with the ocean (Kizielewicz & Luković, 2013). Marine tourism includes activities such as leisure activities, water sports and maritime activities and is gaining popularity (Martínez Vázquez et al., 2021). Recreational travel to coastal and marine areas has probably been around as long as humans. The sea has a strong attraction for people, which is not surprising given its importance as a source of food and transport. As such, marine tourism forms an important component of the wider tourism sector. Especially in many coastal countries, marine tourism has become a primary focus. For these reasons, marine tourism has emerged as the only type of tourism in many cases (Orams, 2002). Marine tourism has a broad structure. Therefore, this branch of tourism, which includes natural, cultural and sportive elements, includes various activities and sub-sectors.

Figure 1. Sub-Tourism Types that can be subject to Maritime Tourism

Source: (Created by the Authors) (Gladstone, Curley & Shokri, 2013).

Cruise Tourism: Some areas of the global tourism industry are growing faster than others and have the potential to generate higher revenues. Cruise tourism is one of these areas (Papathanassis, 2020). It is defined as luxury holiday tours on large cruise ships and cultural and historical excursions in port cities (Nedelcu et al., 2015). Since the 1970s, with the development of air transport and the increase in flight range, the need to attract passengers has led to the popularity of cruise tourism. As cruise companies sought new markets, cruise travelling has become a strong trend in the tourism sector with a wide choice of destinations, themes and different purposes. Modern cruise ships are seen as floating holiday villages with a variety of services and activities. Therefore, these ships have become the primary destination for tourists, while the land destinations they call at are of secondary importance. With the increase in the size of the ships and the variety of services, the effects of cruise tourism have become more and more evident. The continuous increase in the number and size of ships has economic, social and ecological impacts on the destinations visited (Kovačić & Silveira, 2020).

Although cruise tourism is largely centred in North America, the Caribbean is the most important destination. In recent years, Europe as well as Asia and the Pacific regions have become increasingly important. Cruise tourism is of great economic value, especially for island destinations, generating additional tourism revenue through port services fees paid by operators and land-based tourism expenditure by passengers and crew (Kester, 2003). Globally, millions of tourists each year travel across oceans and continents to explore different destinations. In this context, the number of cruise travellers is expected to exceed 35 million in 2027 (CLIA, 2020). While luxury cruises create high economic value (De Melo Rodríguez et al., 2017), they also cause significant emissions and port congestion. Carbon-neutral cruise initiatives are being developed recently to mitigate these negative impacts (Braidotti et al., 2023).

Yacht and Marina Tourism: Mikulić et al. (2015) consider yacht tourism as a type of special interest tourism and include yachting in a broader category of marine tourism. They also use the terms marina tourism and recreational boat tourism as synonyms for yacht tourism. In other words, yacht and marina tourism can be defined as travelling by private yachts and maritime activities served by marinas. Although some studies indicate that yacht tourism faces challenges such as environmental impacts and the effects of the COVID-19 pandemic, it is stated that it has experienced global growth and development due to factors such as luxury consumption, economic benefits, and increased interest in sustainable practices (Stepanets & Hryniuk, 2021; Sevinç & Güzel 2017; Ajagunna & Casanova 2022). However, while yacht tourism represents luxury consumption and niche markets, it also raises sustainability questions. This has led to the emergence of green marina certifications and zero-waste initiatives (Ioras, 2025; Ioras & Bandara 2025).

Beach and Coastal Tourism: Coastal and beach tourism, which is a holiday concept based on the trinity of sun, sand and sea, is a diverse, dynamic, fast-growing and economically attractive form of contemporary tourism that encompasses a variety of tourism, entertainment and leisure-oriented facilities and activities (Lan & Thanh, 2023). Coastal tourism includes beaches as the main resource offering a variety of natural ecosystems, landscapes, and historical and cultural features, such as local gastronomy (Pessoa et al., 2019). Indeed, in this type of tourism, natural attractions are the primary factor in attracting tourists (Rangel-Buitrago et al., 2019).

Some studies suggest that coastal and beach tourism involves activities centred around natural and cultural attractions, with tourist segments motivated by sun, beach and leisure or ecological and cultural experiences, but facing challenges such as environmental degradation, climate change impacts and governance issues that require sustainable management practices (Onofri & Nunez, 2013; Liang & Ding, 2002; Gössling et al., 2018). This type of tourism relies on sunshine, sand and sea, but it is vulnerable to erosion, climate change and pollution. Consequently, coasts and beaches are receiving greater protection in the form of Blue Flag beach designations (Merino & Prats 2022).

Diving and Water Sports Tourism: Diving tourism is a low-impact recreational activity that allows for environmental protection and socioeconomic benefits for local communities (Cavallini et al. 2023). Diving tourism is a rapidly growing tourism activity with a high potential to attract young individuals in all source markets (Kothalawala & Arachchi, 2021). As a matter of fact, it can be stated that it is a special type of tourism where people dive to explore the underwater world, providing income and employment to local communities as well as contributing to marine conservation and sustainability efforts (Dimmock & Musa, 2015). In addition, it contributes at least USD 4 billion annually to marine ecosystems and helps finance marine conservation initiatives and research (Nababan, 2023). Apart from diving tourism, another type of tourism that can be considered within the scope of marine tourism is water sports tourism. This type of tourism includes sailing, cruising, motor boating, surfing, windsurfing, snorkelling and free diving as well as rafting, canoeing and sailing training adventures (Jennings, 2007). Although diving tourism is important for exploring and contributing to the economy of marine ecosystems, it can also have negative impacts, such as physically damaging coral reefs, disturbing marine life and disrupting the ecological balance (Dimopoulos et al., 2025). However, environmentally responsible diving programmes, awareness-raising training and sustainability-focused practices can mitigate these threats and promote marine conservation (Fernández & Rivera 2024).

Fishing and Hunting Tourism: Sport fishing at sea refers to any recreational use of marine fisheries, excluding commercial sale and fishing for profit (Pawson et al., 2008). Despite concerns over food security and overfishing, recreational fishing contributes to human nutrition by providing an accessible, affordable and sustainable food source (Cooke et al., 2018). However, it is important that guided hunts in the seas are carried out in an organised and rule-bound manner. As a matter of fact, in addition to protecting the environment, it seems possible to keep species away from threats by adhering to the rules (Velando & Munilla, 2011).

Health Tourism: The therapeutic properties of the seas are combined with tourism and offer unique opportunities for health, spirituality and sustainable development (Jonas, 2024). As a matter of fact, health tourism activities in the seas can be stated as tourism activities aiming to improve the physical, spiritual and mental conditions of people through the effective use of natural healing resources (Chomka & Gut-Winiarska, 2020; Kim & Boo, 2015).

Thalassotherapy, which is one of the practices within the scope of health tourism at sea, is defined as the use of sea water and products obtained from the sea together for preventive and therapeutic purposes under medical supervision in facilities near the sea (Gomes et al., 2021). It is stated that thalassotherapy provides significant improvements in disease severity and quality of life in skin and rheumatic disorders, and the strongest evidence is seen in the treatment of psoriasis and fibromyalgia (Antonelli & Donelli, 2024). On the other hand, climatotherapy applications are the planned medical application of climatic factors to improve functions and prevent or treat diseases (Amatya & Khan, 2023). As a matter of fact, SPA and health centres established close to the sea provide holistic treatments by using sea water and environments, while offering health and welfare environments to tourists in the seas (Bjelland, 2006).

Marine Ecotourism: The least that can be said about marine ecotourism is that it is a subset of both marine nature-based tourism and sustainable marine tourism (Wilson & Garrod, 2003). Marine ecotourism is an alternative income-generating activity to fishing that relies on marine ecosystems, reefs and protected areas for tourism (Cusack et al., 2021). As a matter of fact, this type of tourism is becoming more important depending on the resources in marine areas and its economic potential is becoming more and more recognised day by day. Apart from its economic potential, it is stated that the marine ecosystem has a significant potential to produce a wider range of benefits. In this context, marine ecosystems can provide benefits such as funding for research on marine species and habitats, protection of marine resources, economic resources for environmental management, and providing a focal point for social and cultural regeneration of coastal communities (Wilson & Garrod, 2003).

Marine ecotourism provides the opportunity to examine the life in the seas more closely. In this context, it can be stated as environmentally friendly activities that include examining coral reefs, observing sea creatures (sea turtles, dolphins, whales), bird watching in coastal wetlands and interacting with local communities (Gladstone et al., 2013). In this direction, when the scope of marine ecotourism is considered, it can be said that it is ecotourism that takes place in marine and coastal environments and can be part of an appropriate strategy to solve the problems encountered in coastal environmental areas (Garrod & Wilson, 2004). Indeed, marine ecotourism not only demonstrates an environmentally sound approach and supports the conservation of natural resources, but also provides alternative income opportunities for local communities. Community-based practices in Mexica and Norway are particularly renowned for their success in this area (De la Cruz-Modino & Cosentino 2022; Cisneros-Montemayor et al., 2020).

Historical Development of Marine Tourism

Recreational travel to the sea and coastal areas has always existed throughout history. Although these areas are visited for fishing and shellfish collection, they have also been important places that have attracted people for recreational activities such as swimming, exploring, resting and social activities for many years. It is even stated that there are records showing that the seas are also preferred for activities such as sailing for pleasure (Orams, 2002).

Coastal areas attract the most tourists and the fastest growth in the tourism sector has been in the coastal and marine tourism sub-division. This area is very broad and diverse, encompassing a wide range of activities on the coastline, in nearshore waters, offshore, underwater and niche activities for specific interests (Gladstone et al., 2013). Travel for leisure, recreation and holidays recovered slightly, reaching 56% of all international arrivals in 2022, after falling to 53% and 54% in the pandemic years 2020 and 2021. In contrast, travel for visiting friends and relatives, health and religious purposes grew in 2020 and 2021, before falling to 29% in 2022 (World Tourism Organization, 2024). It is possible to say that marine and coastal tourism still occupies an important place in the sector, as the purpose of tourists' visits to destinations is mostly for entertainment, recreation and holiday purposes, and the sea and coasts are more suitable areas for these visit purposes. Therefore, marine and coastal tourism is often referred to as one of the dominant sectors in many ocean economies, especially in terms of economic contribution and the number of people employed (Hynes et al., 2024). It can be stated that marine tourism has an impact on regional development, infrastructure and service development, development and promotion of local products, and environmental and cultural awareness of local people, as well as having an impact on a global scale such as diversification of tourism and promotion of cultural and natural heritage. In other words, it can be stated that marine tourism is important in terms of economic growth, sustainable development, ecosystem protection, social participation and tourist satisfaction (Kasim et al., 2021; Li, Wu & Partwary, 2022; Sukran et al., 2025). While marine tourism is a vital sector supporting economic development and employment, it also poses risks of overexploitation of natural resources and environmental degradation. Therefore, maintaining the balance between opportunities and risks is critical. The COVID-19 pandemic has highlighted the fragility of maritime tourism, highlighting the need for more resilient and sustainable recovery strategies (Kolesnikova, 2020).

The Ecological Impacts of Marine Tourism

Tourism is one of the largest and fastest growing sectors in the global economy and has significant environmental, cultural, social and economic impacts, both positive and negative. Tourism can be an important tool for economic development, but if not properly planned, it can have devastating impacts on biodiversity and pristine environments and lead to the misuse of natural resources such as freshwater, forests and marine life (GhulamRabbany et al., 2013). With global tourist arrivals reaching 1.3 billion in 2023 (World Tourism Organization, 2024), it is not easy to determine the consequences of tourism, even if it is assumed that such a large activity has a significant impact on the environment (Gössling, 2002). As a matter of fact, the reasons such as focusing on coastal destinations rather than the marine environment in determining the possible effects of tourism on the seas, the heterogeneous structure of the marine environment, the variability of the effects in the marine environment, and the variability in the speed of the effects in different regions can be stated as difficulties in determining the effects of tourism (Swarbrooke, 2020). For this reason, it would be a more accurate approach to address some of the prominent results of the ecological impacts caused by marine tourism.

Impacts on Marine Ecosystems

As is well known, tourism can contribute economically, socially and environmentally locally, regionally and nationally. In other words, tourism can contribute financially to conservation, but more importantly, it can provide interest and awareness for conservation and act as a catalyst for the conservation of resources (McKegg, 1999). The positive effects of tourism on marine areas can be shown by carrying out tourism activities based on the understanding of sustainability of living and non-living assets in the sea. As a matter of fact, studies show that tourism can contribute positively to the protection of marine ecosystems, creation of economic value and sustainability of ecosystem services through sustainable tourism practices and conservation efforts (Prihadi et al., 2024; Bartolini et al., 2024; Lopez-Rivas & Cardenas, 2024; Bryhn et al., 2020; Otrachshenko & Bosello, 2017).

Carrying out marine tourism activities with an understanding that prioritises economic growth can also have negative impacts on marine ecosystems. In other words, there may be pressure or threats on living beings in the seas due to over-commercialised tourism activities. The impacts caused by tourism activities can be seen especially on marine reptiles, seabirds, marine mammals and coral reefs (McKegg, 1999). In this context, Avila et al. (2018) reported that human activities in coastal waters worldwide pose significant cumulative risks to most marine mammal species and that 51% of the core habitats of these species are high-risk areas.

Bearzi (2017) states that although marine mammal-based ecotourism has economic benefits and supports local economies, there are concerns about the sustainability of the popularity of this type of tourism and its potential negative impacts on marine species. Culhane et al. (2024) state that fishing and tourism and recreation activities pose the greatest risk to tropical marine ecosystems, with coral reefs being the most vulnerable. Gooden et al. (2024) found that cage diving boats using food-based attractants increased white shark activity levels, minimizing changes in energy levels and potentially affecting their health and fitness. Some studies have also shown that marine tourism can lead to behavioural changes, health problems and ecological degradation in marine reptiles, and that plastic pollution and anthropogenic wastes can cause serious damage such as entanglement and ingestion, especially in species such as sea turtles (Burgin & Hardiman, 2015; Bottari et al., 2024; Staffieri et al., 2019). Indeed, research indicates that marine tourism can have an impact on the marine ecosystem. However, shortcomings in conservation efforts are also a noteworthy factor in addition to these negative pressures (Dong, 2025).

Degradation of Coastal Areas and Erosion

While economic growth in coastal areas comes to the fore as a result of marine tourism activities (Liu & Cao, 2018), the increase in unsustainable practices can also bring along various environmental problems (Swarbrooke, 2020). As a matter of fact, some studies show that in coastal areas where this economic growth is experienced but not properly managed, negative effects such as environmental pollution, habitat degradation and waste increase can occur (Ji & Ding, 2024; Machado & de Andrés, 2023; Miller & Auyong, 1991; Swarbrooke, 2020).

As marine tourism activities take place in the seas, various impacts can also be seen in coastal areas due to the fact that the sea and coasts are inseparable parts. In this context, problems such as deterioration of beach ecosystems in coastal areas (Buzzi et al., 2022), metal pollution on beaches due to overdevelopment of coasts (Vetrimurugan et al., 2017), direct disposal of garbage in the beach environment (Maione, 2021). In this context, Dal and Baysan (2007) stated in their study that the Kuşadası coasts in Turkey experience intensive tourism activities in the summer months and as a result, the pressure experienced by the coastal areas causes environmental and ecological degradation.

On the other hand, coastal erosion also increases in coastal areas due to the development of marine tourism. Buildings, breakwaters and sea embankments built on the coast, artificial widening of beaches, destruction of vegetation and overuse of the coast can be stated as the reasons that trigger erosion (Swarbrooke, 2020; Razak et al., 2023). In this direction, Görmüş et al. (2014) conducted a study on the touristic Karasu coast of Sakarya province and observed that harbour structures increase coastal erosion. Between 1987 and 2003, it was observed that the coastal erosion was approximately 100 m. As a result of the study, it was understood that the harbour constructed in the region had a great effect on this erosion. The coasts are an important point for the tourism sector in Turkey and although coastal cities cover less than 5 per cent of the total surface area of Turkey, they account for 51 per cent of the population and 90 per cent of tourism revenues. The erosion of Turkey's coasts results in the loss of about 6 per cent of GDP for capital loss (Albayrakoğlu, 2011). However, unplanned and inadequately developed infrastructure practices threaten ecological sustainability by causing accelerated coastal erosion not only in Turkey but also on the African coasts (Charuka et al., 2023). On the other hand, the degradation and erosion of the coasts as a result of the uncontrolled development of marine tourism can be environmental and social, as well as economic losses. As a matter of fact, Garcés-Ordóñez et al. (2020) stated that the reduction of negative ecological and socioeconomic impacts can be achieved through stronger controls, education and awareness strategies.

Water Pollution and Waste Management Issues

Marine tourism is a sector in which water plays a central role both naturally and economically. The quality, accessibility and sustainable management of water are critical to the success and sustainability of marine tourism. Marine tourism is directly related to the physical properties of water (cleanliness, clarity, mobility) (Rhoden & Kaaristo, 2020). Indeed, the quality and accessibility of water affect both the satisfaction of tourists and the quality of life of local people, especially on islands and in water-dependent regions (Kim et al., 2024).

On the other hand, protection and management of water resources is an issue that needs to be addressed in terms of economic growth and sustainability for marine tourism (Li et al., 2022). Deterioration in sea water quality causes tourists to decrease their desire to repeat and thus negatively affects the economic income of destinations (Schuhmann et al., 2019). For these reasons, protection of water quality in coastal areas is necessary for ecosystem health and sustainability of tourism activities. Poor water management, pollution and excessive tourism reduce water quality, threatening both biodiversity and tourism potential (Dreizis, 2020; Kurniawan et al., 2023).

While the water being different from normal affects marine tourism, uncontrolled marine tourism also causes water pollution and some problems to arise as a result. Among the different effects of marine tourism, its effect on public health is important. Indeed, in touristic destinations that already have scarce resources, the increase in the consumption of water and especially fresh water resources due to the periodic increase in the population generally causes pollution of the sea and serious problems. As a result of excessive consumption, numerous sewage discharges affect coastal water quality. Accommodation facilities serving in coastal areas discharge the wastewater generated into the sea without any treatment, causing the water to deteriorate. In addition to accommodation facilities, water resources are overused for swimming pools, golf courses and personal water use of tourists. This situation causes water scarcity and deterioration of water resources, and therefore the emergence of more wastewater. The problem of water pollution in the seas can also affect tourists. It can cause various health problems in individuals, especially as a result of exposure to polluted water (Gedik & Mugan-Ertugral, 2019).

The vehicles used within the scope of marine tourism also cause water pollution. In addition to the negative effects of fuel consumption of cruise ships, it is known that the bilge water they release into the sea causes various problems. The fact that this waste water contains various chemicals is another factor that threatens the life in the sea (Brida & Zapata, 2010). On the other hand, the same waste water problem is also seen from yachts and sailing boats. In particular, the uncontrolled discharge of bilge water into the sea causes pollutants such as petroleum derivatives, detergent residues and various toxic substances to mix into the water environment, thus disrupting the chemical balance of aquatic ecosystems and negatively affecting the life of many organisms from plankton to fish (Amin et al., 2024). Eliminating these problems is another problem. It can be said that the deficiencies in the solution of water pollution problems caused by marine tourism are inadequate waste management and deficiencies in policy and community participation. In the case of inadequate waste management, wastes easily reach the sea due to lack of infrastructure and ineffective waste collection systems (Tsai et al., 2021; Łapko et al., 2018; Hayati et al., 2020; Nguyen et al., 2025; Schneider et al., 2018). In terms of deficiencies in policy and community participation, it is seen that there are insufficient legal regulations for sustainable waste management, and social awareness and stakeholder participation are not at a sufficient level (Musyafah et al., 2025; Tsai et al., 2021; Łapko et al., 2018; Hayati et al., 2020; Nguyen et al., 2025). As a matter of fact, it can be said that integrated management is needed to solve the problems. In this context, for the sustainability of marine tourism, waste management, water quality monitoring and dissemination of environmentally friendly practices (Zheng and Liu, 2021; Gössling et al., 2018; Mejjad et al., 2023; Kurniawan et al., 2023) and legal and political measures should be taken (Zheng & Liu, 2021; Mejjad el al., 2023).

Relationship Between Climate Change and Marine Tourism

Marine tourism interacts with climate change. Marine tourism is both affected by climate change and can indirectly increase climate change through the pressures it creates in coastal areas. The effects of marine tourism on climate change are related to the pressures on coastal ecosystems, the use of infrastructure and the increase in local environmental stress factors.

Marine tourism can increase the impacts of climate change by putting pressure on coastal and marine ecosystems. Local environmental stressors and infrastructure pressures increase vulnerability to climate change. Marine tourism, together with the expansion of coastal infrastructure, destruction of natural areas and increased human activity, creates stress in coastal ecosystems. This stress can increase the impacts of climate change and lead to losses in ecosystem services. (Weatherdon et al., 2016; Arabadzhyan et al., 2021; Gissi et al., 2021). Local stressors such as pollution, waste generation and habitat degradation associated with marine tourism can exacerbate the impacts of climate change on the marine environment. This increases vulnerability to climate change, especially in sensitive coastal areas (Van Putten et al., 2014; Gissi et al., 2021). On the other hand, increased tourism activities can increase nutrient loads and pollution, triggering harmful algal blooms. With climate change, the frequency and severity of these events are increasing, which leads to negative effects on tourism, fisheries and human health (Gobler, 2020). The impact of maritime tourism on climate change can be given as an example through cruise tourism. The high fuel consumption of cruise ships constitutes a large part of the greenhouse gas emissions originating from transportation by sea. It is known that the use of fossil-based energy, especially the intensive consumption of diesel-based heavy fuels, significantly increases harmful emissions such as carbon dioxide (CO_2), sulfur oxides (SOx) and nitrogen oxides (NOx); this situation accelerates global climate change by triggering atmospheric pollution and acidic sedimentation processes (Corbett & Winebrake, 2008). For example, a single large cruise ship can emit as much carbon dioxide in a day as 80,000 cars (OECD, 2016). This effect deteriorates the air quality, especially in coastal cities, and threatens general public health. As can be understood from the example, it is possible to say that the consumption of large amounts of energy in marine areas has an impact on climate change by changing the greenhouse effect in the atmosphere (Ji & Ding, 2024).

Changes in marine ecosystems can also lead to significant shifts in tourism flows and coastal infrastructure. These impacts mean both risks and new opportunities. Indeed, in some regions, climate change can create new tourism opportunities. For example, warmer climates can make some destinations more attractive (Weatherdon et al., 2016; Coles, 2020). Nevertheless, there are various ways to mitigate the impact of marine tourism on climate change. In this context, it is vital to implement adaptation measures such as renewable energy, coastal protection and resilience planning (Zentner et al., 2023; Roberts et al., 2017).

Social Impacts of Marine Tourism

The economic advantages of marine tourism have been demonstrated to be of significance. However, it would be remiss to address the issue solely from this perspective. The repercussions of this particular form of tourism on societal entities are of paramount importance and warrant meticulous examination. Local inhabitants in coastal regions are subject to the direct impact of marine tourism. The following changes in the culture, demographic structure and lifestyle of the local society have been observed in the wake of marine tourism: The social impacts of tourist activities may vary depending on the intensity, type and duration of the activity. The manner in which tourists and local people interact is indicative of the direction of social impacts. Intensive and direct interactions have been shown to increase cultural exchange and mutual understanding (Ginting & Jayanti, 2024). However, unplanned and excessive density can lead to social tensions or cultural disruptions (Stylidis, 2020). Consequently, it is imperative to incorporate the social dimensions of marine tourism into the development of a sustainable tourism management approach.

The positive social effects of tourism, such as the development of service infrastructures, the creation of employment opportunities for local people, and the enhancement of income levels, thereby affecting overall welfare, are evident in regions where tourism is prevalent. In particular, the economic opportunities engendered by these industries have been demonstrated to encourage reverse migration and to maintain a youthful population in coastal areas (Faulkner & Tideswell, 1997). Furthermore, the development of destinations with regard to transportation, health and education facilitates easier access to such services for local people. Consequently, it also contributes to an enhancement in the quality of life in the region. However, in instances where local populations are unable to benefit from these services and provisions equally, it may also result in the emergence of social inequalities (Bimonte & Faralla, 2016).

Excessive and uncontrolled development of maritime tourism has the potential to result in deterioration in the social structure. An increase in land prices in coastal areas has the potential to result in elevated living costs for the local population and to engender social exclusion. In certain instances, the advent of tourism-oriented spatial changes can result in the marginalisation of traditional local life practices, with artificial cultural representations being constructed according to tourist expectations assuming primacy in these areas (Doxey, 1975). The phenomenon of "cultural alienation", defined as the process by which individuals move away from their own cultural values and adopt those of other cultures, is a salient feature of such cases.

The influx of tourists to the region, precipitated by the advent of maritime tourism, has also given rise to various social problems that have a detrimental effect on social life. This phenomenon is particularly pronounced in destinations that experience a surge in tourist arrivals during the summer season. The consequences of this overcrowding include increased traffic congestion, intensive utilisation of basic resources such as water and energy, and a deterioration in the daily living conditions of the local population (Gössling, 2002). Concurrently, cultural divergences experienced by tourists have the potential to precipitate value conflicts and social tensions (Ayaz & Parlak, 2019).

Sociocultural Impacts on Local People

The impact of marine tourism on cultural change in coastal communities has been a subject of considerable interest. Local inhabitants of coastal areas, which experience heightened levels of tourism during the summer months, observe significant transformations in their social relations and daily lifestyles. It is evident that a number of these transformations are experienced in relation to access to employment opportunities and infrastructure/superstructure opportunities. It is evident that the management of the socio-cultural effects of tourism is of significance not only from the perspective of the local populace but also in terms of the formulation and preservation of a sustainable policy for the region (Ap & Crompton, 1998).

The provision of cultural heritage elements in coastal areas for tourism purposes has been shown to have both positive and negative consequences. In addition to the protection of these areas, their transfer to future generations and the creation of economic value, negative effects such as cultural degeneration, loss of authenticity and physical wear are also encountered due to excessive tourism (Greenwood, 1989).

It is evident that local communities should be the focal point of any strategies devised for the management of cultural heritage, and that a participatory approach should be formulated with these communities. It is imperative that action is taken for two principal reasons. Firstly, in order to protect cultural values, and secondly, so that these values may be experienced and maintained. The approach formulated by UNESCO within the framework of this principle is expressed as the "Living Cultural Heritage" approach (UNESCO, 2003). Local communities should be integrated into the process not only as an element involved in the realisation of cultural elements, but also as the entities responsible for the transmission of this culture to future generations. In light of these findings, it is recommended that local communities be empowered to participate in the decision-making process concerning cultural activities in coastal regions where marine tourism is prevalent (Timothy, 2006).

A series of policies must be implemented to reduce or eliminate the negative socio-cultural impacts originating from coastal tourism. Of these, applications concerning carrying capacity are of particular significance. It is imperative to take precautionary measures regarding this issue, particularly in the context of cultural heritage areas. The implementation of educational programs and projects aimed at raising awareness among local communities is also a policy that can be implemented (McKercher & Du Cros, 2002). Furthermore, cultural assets should not be considered exclusively from an economic perspective, but should also be protected and maintained in their symbolic meanings. It is imperative that young people engage more actively in tourism and cultural activities in order to ensure a sustainable future. The provision of support for the productions of female producers at a local level, the archiving of cultural heritage elements in digital media, and the establishment of a management approach that incorporates all stakeholders can only be achieved by following sustainable policies in a cultural sense. This approach is predicated on the premise that cultural values will be secured by strengthening the bond between individuals in a social sense (Timothy & Boyd, 2003).

Changes in Social Structure

The most significant factor contributing to the substantial social transformations observed in coastal regions is marine tourism. This phenomenon is particularly evident in regions that are heavily reliant on tourism, where the population density tends to rise. Consequently, substantial shifts are evident in the demographic structure and social relations (Papathanassis, 2023). The employment opportunities of the region that develop with tourism also bring a wave of migration. With the serious migrations, a complex and multicultural social structure is formed, thus many coastal settlements with rural elements are faced with the phenomenon of urbanization (Williams & Hall, 2000).

The economic disparities within society are exacerbated by the high cost of marine tourism. The increase in investments in beach clubs, cruise ports and luxury yacht marinas that appeal to high-income people has had an economic impact on local communities and tourists. It is evident that the rise in the value of land and plots within the region has a direct impact on the economic well-being of local inhabitants, particularly those who are financially disadvantaged. This phenomenon results in a migration of these individuals from the area, as they seek opportunities elsewhere. This phenomenon gives rise to a perception of social injustice, thereby disrupting the prevailing social structure (Bianchi, 2003). The tendency for tourism investments, which have been shown to yield high economic returns, to be directed towards large capital owners has the potential to exacerbate perceived inequalities within society (Castilho & Fuinhas, 2025).

The impact of marine tourism on the social structure is twofold. Firstly, it ensures the participation of women in the workforce. However, as Kabil et al. (2024) demonstrate, gender inequality may arise when the female workforce is predominantly employed in roles such as kitchen work, cleaning and sales. The attainment of economic independence by women engenders a shift in the balance of power within the household, thereby facilitating a re-evaluation of established gender roles.

The impact of tourism on social structures can also result in substantial transformations within the family institution. The transition from the traditional extended family structure to the nuclear family model is becoming increasingly rapid, particularly in the context of young individuals entering the tourism sector and achieving economic independence at an early age. Moreover, the increase in seasonal and migrant labour has been shown to engender phenomena such as the propagation of extramarital unions, the escalation in divorce rates, and the redefinition of family roles (Sharpley, 2014). In this context, it can be posited that marine tourism engenders a multi-layered effect that has the capacity to transform the social fabric of societies.

Impacts on Income Distribution and Employment

Maritime tourism is a notable phenomenon in this regard, given its capacity to stimulate economic dynamism in coastal regions. Nevertheless, the repercussions of this growth on local communities are frequently the subject of critical assessments, particularly with regard to the spatial and socioeconomic distribution of benefits (Hynes et al., 2024). Investments in tourism-oriented enterprises in small-scale settlements have the potential to generate employment opportunities in both direct and indirect capacities. This phenomenon can be viewed as favourable from the perspective of regional development policies. However, the question of whether the employment and economic returns in question reach all segments of society in an equitable manner constitutes a problem that cannot be ignored in terms of the long-term sustainability of the tourism sector (Ashley & Roe, 2001). In coastal cities, where the local population is predominantly engaged in low-income and seasonal employment, it has been observed that tourism-related revenues are disproportionately retained by multinational tourism companies and external investors (Nguyen et al., 2020).

When evaluated in terms of its capacity to generate employment opportunities, marine tourism has been found to engender seasonal job prospects (Pham, 2020). Marine tourism, which has been demonstrated to facilitate the participation of women and young people in the workforce, is a significant social benefit in this regard. However, the sector's employment prospects are characterised by limited financial rewards, restricted career progression and an absence of social security benefits, thus highlighting its significant drawbacks (Baum, 2007). The evidence suggests that the employment of migrants, children and women in the informal sector, or with flexible working hours and opportunities, serves to perpetuate social inequality. As marine tourism provides intensive seasonal employment opportunities, local people may become unemployed during the off-season (Radlińska & Gardziejewska, 2022).

The consequences of tourism activities on income distribution are frequently asymmetrical. While certain actors – especially those with substantial capital capacity – reap considerable economic benefits from tourism, a considerable proportion of the local population can only benefit from this growth to a limited extent (Gilliland et al., 2020). A significant contributing factor to this imbalance is the prevalence of firms operating in high-value-added service sectors, such as qualified accommodation facilities, yacht tourism, and underwater sports, that are associated with external capital. This phenomenon, conceptualised as the "leakage effect" in the extant literature, manifests itself in the form of a significant proportion of the income generated in the destination being transferred to external investors or main centres instead of the local economy (Telfer & Sharpley, 2015). Consequently, in order for marine tourism to contribute to the regional development of the region, an inclusive economic strategy plan that takes local capacity into account should be developed. The promotion of cooperatives, the support of local suppliers, and the encouragement of small-scale businesses have been identified as key strategies for achieving a more equitable distribution of the total economic benefit from tourism (Torres & Momsen, 2004). The integration of the local population into a qualified workforce can be achieved through education and vocational qualification programmes structured to support this process. Such programmes can result in a quantitative increase in employment and a quality-oriented transformation in the labour market. This, in turn, will contribute to the increase in social living standards in the long term (Danlu, 2023).

Effects on Cultural Heritage and Identity

It would be erroneous to discuss the relationship between only the natural environment and marine tourism in coastal settlements affected by marine tourism. Furthermore, marine tourism has been shown to be closely related to the cultural values and identities of the local population (Baihaqi & Annida, 2024). Activities undertaken within the domain of marine tourism encompass historical elements, cultural practices, and traditional motifs in coastal regions, presenting them in the context of contemporary conditions. The integration of cultural heritage elements into the domain of tourism activities is a strategy that has been proven to ensure the preservation of cultural heritage. Conversely, it has been argued that this can result in the transformation of these elements into commercial entities, thereby leading to the dilution of their original significance (Greenwood, 1989).

In regions where tourism is prevalent, cultural heritage elements are increasingly being detached from their original contexts and reshaped in accordance with external demands. This phenomenon gives rise to the notion that external observers assume a pivotal role in the construction of local identities. When evaluated in the context of the conceptualisation of "staged authenticity" introduced to the literature by MacCannell (1976), authentic life practices are adapted to the taste of visitors; thus, cultural expressions are reduced to a superficial and often commercial representation. In this transformation process, traditional art forms, rituals, maritime traditions and other local practices become divorced from their original worlds of meaning, evolving into a mere visual and experiential consumption object. The present generation of young people is acutely impacted by the ongoing process of cultural identity renewal. It is an established fact that young people who engage with citizens of other countries through the medium of tourism are afforded an opportunity to compare their own culture with that of other cultures. Consequently, their sense of belonging to their culture may be reinforced, or they may experience a sense of alienation. Moreover, the pervasive utilisation of foreign languages in tourist destinations has the potential to inflict harm upon the indigenous linguistic traditions of the region (Smith, 2015). When evaluated within this framework, the phenomenon of tourism offers both transformation opportunities and potential threats that may lead to the damage of cultural integrity in the process of shaping the collective identities of local communities. The redefining of identity through representations aimed at visitors, on the one hand, increases socio-cultural visibility; on the other hand, however, it may pose the risk of the instrumentalising of authentic elements (Wang & Zhang, 2020).

The management of cultural heritage, when organised within an inclusive and sustainable framework, has been demonstrated to positively impact tourism's effect on cultural identity. In the evaluation of cultural heritage, it is essential to consider not only economic returns but also historical continuity, symbolic meaning and social belonging dimensions. In the process of heritage protection, it is of great importance that local people are positioned not as passive spectators but as the primary subjects of cultural production and transmission (Moric et al., 2021). In this regard, following the approach of the United Nations Educational, Social and Cultural Organization (UNESCO) to the concept of "intangible cultural heritage," there is a necessity for the development of sustainable policies in collaboration with local actors, with the aim of preserving traditional knowledge systems, ceremonial practices, narrative types and handicraft forms. Furthermore, educational initiatives that enhance cultural capacity should be supported (UNESCO, 2003). Such holistic approaches enable cultural identity to develop resistance against external effects and to maintain its existence in the long term.

Tourism and Social Conflicts

Although tourism is often considered as an element that promotes socio-economic development and facilitates intercultural communication, it can create various tension areas that can trigger social unrest, especially in regions where it develops without control mechanisms (Yang et al., 2013). Touristic mobility, especially observed in coastal areas where marine tourism is concentrated, paves the way for increasingly deepening conflicts of interest between the life practices of local people, tourist demands and the economic goals of investors. In the background of such conflicts, structural factors such as imbalances in access to economic resources, incompatibility between cultural value systems, the pressure of spatial transformation processes on local identity and increasing demand for the natural environment play a decisive role (Mbaiwa, 2005).

The daily lives of local people in regions affected by tourism are directly impacted by tourism activities. It is evident that the transformation of areas inhabited by local residents for tourism purposes engenders a profound shift in the prevailing social order within the region. The restructuring of public settlements through various means, including urban transformation for the construction of tourism facilities, has been demonstrated to result in population migration (Dogan, 1989). In regions where this situation, termed “displacement”, is experienced, social unrest is likely to occur. In regions where tourism-based facilities are dense, coastal areas are allocated to private companies, which has the effect of restricting local people from benefiting from the sea. This has the potential to lead to social unrest (Unhasuta et al., 2021).

In regions where coastal tourism is concentrated, conflicts between cultural norms and social value systems can trigger various socio-cultural tensions between local communities and tourists. In particular, visitors' behaviour in public spaces, dressing styles, alcohol consumption or insensitive approaches to privacy perceptions can be perceived as a cultural discomfort element by communities with traditional values. Such normative conflicts can be interpreted in the context of the “Irridex Model” developed by Doxey (1975). According to the relevant model, although host communities initially approach tourism activities with tolerance and support, they may tend to develop reactions such as anger, tension and open opposition with increasing dissatisfaction over time. As this process progresses, there is the possibility of a weakening of social solidarity, a disturbance of the perception of security, and the observation of tendencies towards social polarisation. Incorporating local inhabitants into the tourism planning process is instrumental in ensuring that tourism-related unrest within the community is mitigated. In the context of a tourism planning process that engages the public, it is anticipated that the public will embrace tourism and support its sustainability (Tosun, 2000). Furthermore, the implementation of cultural protection policies aimed at strengthening social equality, the adoption of holistic economic strategies to eliminate imbalances in income distribution, and awareness-raising activities aimed at providing cultural sensitivity to visitors can contribute to the reduction of potential cultural frictions (Thi et al., 2024). Conversely, structural developments in the domain of marine tourism have the potential to act as a catalyst for the fragmentation of the prevailing social structure, thereby exacerbating the existing disparities between communities. This phenomenon stands in contrast to the anticipated role of marine tourism as a catalyst for interaction, which is expected to foster local development (Hall, 2001).

Governance, Policy, And Technological Innovation in Marine Tourism

Policy Frameworks: Strong policy frameworks are necessary to balance the social and ecological impacts of marine tourism. Carrying capacity limits and designated coastal zones help to protect natural resources and improve the quality of life for local communities (Saveriades, 2000). Eco-certification systems encourage tourism businesses to fulfil their environmental responsibilities (Font & Buckley, 2001), while international agreements and collaborations contribute to the global protection of marine ecosystems (UNEP, 2006). Examples of regional-level integration of sustainable development with ecosystem protection can be seen in the European Union's Blue Growth Strategy and coastal policies implemented along African coasts (Chen et al., 2020; Kaczynski, 2011).

Case Studies: Various regions have implemented concrete examples that demonstrate successful approaches to managing the social and ecological impacts of marine tourism. For instance, the quota system introduced on Australia's great barrier reef helps prevent excessive tourist pressure (Davis & Tisdell, 1996), while Norway's cruise ship emissions taxes contribute to protecting air quality and promoting public health (Simonsen et al., 2019). In Latin America, coral reef protection programmes ensure the sustainability of ecosystems and create alternative income sources for local communities (Spalding et al., 2017).

Technological innovations: Technological innovations offer significant opportunities for monitoring and managing the social and ecological impacts of marine tourism. Remote sensing systems and artificial intelligence-based monitoring tools allow us to assess the condition of coastal ecosystems and detect environmental threats early on (Jambeck et al., 2015; Kerr & Ostrovsky 2003). Digital participation platforms strengthen social sustainability by actively involving local people in decision-making processes (Bramwell & Lane 2011). Conversely, smart port management systems reduce emissions and minimise environmental impacts by controlling overcrowding in ports (Acciaro et al., 2014).

CONCLUSION

Marine tourism is a rapidly growing tourism branch that includes various recreational and touristic activities in marine and coastal areas. Marine tourism includes a wide range of activities such as beach tourism, boat tours, water sports, diving, whale watching and provides significant contributions both economically and socially. In this respect, marine tourism has attracted the attention of researchers and research on this subject has increased in the last thirty years and integrated approaches that take into account environmental, economic and social dimensions have been put forward (Guo et al., 2024). Indeed, the prominent effects of marine tourism in ecological and social dimensions have been addressed in this research.

Marine tourism has an important place in the economies of many countries. Its main effects are creating employment and contributing to local economies. It is seen as a driving force, especially in developing countries, as the blue economy. However, sustainable management is of critical importance in terms of reducing environmental impacts and protecting natural resources. However, prioritizing the concept of economic growth and overly commercialized touristic activities bring problems in marine areas. Marine tourism activities not only cause changes in the culture and living conditions of the local people in terms of social aspects, but also can have effects on the coasts, waters, biodiversity and climate in terms of ecology. When these effects emerge as a negative situation, they become a complex problem that is difficult to manage. Because marine tourism activities cover a very wide area and naturally even a small mistake that is not taken into account can make the problem much bigger. Therefore, an action based on strategic planning is required to prevent the emergence of negative effects of marine tourism or to eliminate problems. In this context, it is most important for all stakeholders of tourism (local people, tourists, central and local governments, tourism businesses, NGOs, educational institutions) to act within a common understanding. Indeed, it is difficult to manage the ecological and social problems of marine tourism, but it is critical that all stakeholders act within the framework of equal responsibility instead of a single stakeholder in the management of these problems. The following suggestions can be made within the scope of the research:

1. Balance the ecological and social dimensions.
2. Implement participatory governance frameworks.
3. Use technology to monitor and mitigate.
4. Promote equitable distribution of benefits.
5. Strengthen climate change adaptation and resilience planning.
6. Guide the post-Covid recovery process using sustainability principles.

REFERENCES

Acciaro, M., Vanelslander, T., Sys, C., Ferrari, C., Roumboutsos, A., Giuliano, G., Lam, J. S. L., & Kapros, S. (2014). Environmental sustainability in seaports: A framework for successful innovation. *Maritime Policy & Management, 41*(5), 480–500. DOI: 10.1080/03088839.2014.932926

Ajagunna, I., & Casanova, S. (2022). An analysis of the post-COVID-19 cruise industry: Could this be a new possibility for the luxury yacht sector in the Caribbean? *Worldwide Hospitality and Tourism Themes, 14*(2), 115–123. DOI: 10.1108/WHATT-12-2021-0158

Albayrakoğlu, E. P. (2011). Climate change and security: The case for Turkey. *Gazi Akademik Bakış, 5*(9), 59–75.

Amatya, B., & Khan, F. (2023). The efficacy of climatotherapy on medical rehabilitation outcomes: A systematic review. *The Journal of the International Society of Physical and Rehabilitation Medicine, 6*(3), 47–59. DOI: 10.1097/ph9.0000000000000011

Amin, A. D. N., Rashdan, U., Mahadi, W. N. F. H. W., & Abdul, M. H. (2024). Environmental Injustice at Sea: The Impacts of Bilge Dumping on Indonesia's Coastal Communities. *International Journal of Research and Innovation in Social Science, 8*(12), 3559–3568.

Antonelli, M., & Donelli, D. (2024). A narrative review of thalassotherapy and the health benefits of seawater and coastal climates. *International Journal of Biometeorology*, •••, 1–10.

Ap, J., & Crompton, J. L. (1998). Developing and testing a tourism impact scale. *Journal of Travel Research, 37*(2), 120–130. DOI: 10.1177/004728759803700203

Arabadzhyan, A., Figini, P., García, C., González, M. M., Lam-González, Y. E., & León, C. J. (2021). Climate change, coastal tourism, and impact chains–a literature review. *Current Issues in Tourism, 24*(16), 2233–2268. DOI: 10.1080/13683500.2020.1825351

Ashley, C., & Roe, D. (2001). Pro-poor tourism strategies: Making tourism work for the poor: A review of experience.

Avila, I. C., Kaschner, K., & Dormann, C. F. (2018). Current global risks to marine mammals: Taking stock of the threats. *Biological Conservation, 221*, 44–58. DOI: 10.1016/j.biocon.2018.02.021

Baihaqi, F., & Annida, S. B. (2024). The Social and Cultural Impact of Marine Tourism on Coastal Communities: A Qualitative Study at Tanjung Lesung Beach.

Bartolini, A., Di Gennaro, V., Reas, V., Mascolo, R. A., La Notte, A., Capriolo, A., & Ferrini, S. (2024). Ecosystem accounting for marine-based tourism provided by Posidonia oceanica in Italy. *One Ecosystem, 9*, e129751. DOI: 10.3897/oneeco.9.e129751

Baum, T. (2007). Human resources in tourism: Still waiting for change. *Tourism Management, 28*(6), 1383–1399. DOI: 10.1016/j.tourman.2007.04.005

Bearzi, M. (2017). Impacts of marine mammal tourism. *Ecotourism's Promise and Peril: A Biological Evaluation*, 73-96.

Bianchi, R. V. (2003). Place and power in tourism development: Tracing the complex articulations of community and locality. *Pasos (El Sauzal), 1*(1), 13–32. DOI: 10.25145/j.pasos.2003.01.002

Bimonte, S., & Faralla, V. (2016). Does residents' perceived life satisfaction vary with tourist season? A two-step survey in a Mediterranean destination. *Tourism Management, 55*, 199–208. DOI: 10.1016/j.tourman.2016.02.011

Bjelland, A. K. (2006). Selje Spa Thalasso-en helse-og velværekatedral?. *Tidsskrift for kulturforskning, 5*(2).

Bottari, T., Mghili, B., Gunasekaran, K., & Mancuso, M. (2024). Impact of plastic pollution on marine biodiversity in Italy. *Water (Basel), 16*(4), 519. DOI: 10.3390/w16040519

Braidotti, L., Bertagna, S., Rappoccio, R., Utzeri, S., Bucci, V., & Marinò, A. (2023). On the inconsistency and revision of Carbon Intensity Indicator for cruise ships. *Transportation Research Part D, Transport and Environment, 118*, 103662. DOI: 10.1016/j.trd.2023.103662

Bramwell, B., & Lane, B. (2011). Critical research on the governance of tourism and sustainability. *Journal of Sustainable Tourism, 19*(4-5), 411–421. DOI: 10.1080/09669582.2011.580586

Brida, J. G., & Zapata, S. (2010). Cruise tourism: Economic, socio-cultural and environmental impacts. *International Journal of Leisure and Tourism Marketing, 1*(3), 205–226. DOI: 10.1504/IJLTM.2010.029585

Bryhn, A., Kraufvelin, P., Bergström, U., Vretborn, M., & Bergström, L. (2020). A model for disentangling dependencies and impacts among human activities and marine ecosystem services. *Environmental Management*, *65*(5), 575–586. DOI: 10.1007/s00267-020-01260-1

Burgin, S., & Hardiman, N. (2015). Effects of non-consumptive wildlife-oriented tourism on marine species and prospects for their sustainable management. *Journal of Environmental Management*, *151*, 210–220. DOI: 10.1016/j.jenvman.2014.12.018

Butler, R. (2006). The future and the TALC. *The tourism area life cycle: Conceptual and theoretical issues, 2*, 281-290.

Buzzi, N. S., Menendez, M. C., Truchet, D. M., Delgado, A. L., & Severini, M. F. (2022). An overview on metal pollution on touristic sandy beaches: Is the COVID-19 pandemic an opportunity to improve coastal management? *Marine Pollution Bulletin*, *174*, 113275. DOI: 10.1016/j.marpolbul.2021.113275

Castilho, D., & Fuinhas, J. A. (2025). Exploring the effects of tourism capital investment on income inequality and poverty in the European Union countries. *Journal of Economic Structures*, *14*(1), 6. DOI: 10.1186/s40008-025-00349-2

Cavallini, I., Marzo, D., Scaccia, L., Scipioni, S., & Niccolini, F. (2023). Scuba diving tourism and the challenge of sustainability: Evidence from an explorative study in North African-Mediterranean countries. *EuroMed Journal of Business*, *20*(5), 1–26. DOI: 10.1108/EMJB-04-2022-0085

Charuka, B., Angnuureng, D. B., & Agblorti, S. K. (2023). Mapping and assessment of coastal infrastructure for adaptation to coastal erosion along the coast of Ghana. *Anthropocene Coasts*, *6*(1), 11. DOI: 10.1007/s44218-023-00026-6

Chen, S., De Bruyne, C., & Bollempalli, M. (2020). Blue economy: Community case studies addressing the poverty–environment nexus in ocean and coastal management. *Sustainability (Basel)*, *12*(11), 4654. DOI: 10.3390/su12114654

Chomka, M., & Gut-Winiarska, M. (2020). Prozdrowotne aspekty turystyki w środowisku morskim. Nauka i Społeczeństwo. *Przegląd nauk społecznych, technicznych, medycznych i pedagogicznych*, 1(1), 103-112.

Cisneros-Montemayor, A. M., Becerril-García, E. E., Berdeja-Zavala, O., & Ayala-Bocos, A. (2020). Shark ecotourism in Mexico: Scientific research, conservation, and contribution to a Blue Economy. [). Academic Press.]. *Advances in Marine Biology*, *85*(1), 71–92.

CLIA (Cruise Lines International Association). (2020). *2020 State of the Cruise Industry Outlook.* https://cruising.org

Coles, T. (2020). Impacts of climate change on tourism and marine recreation. *MCCIP Science Review*, •••, 593–615. DOI: 10.14465/2020.arc25.tre

Cooke, S. J., Twardek, W. M., Lennox, R. J., Zolderdo, A. J., Bower, S. D., Gutowsky, L. F., Danylchuk, A. J., Arlinghaus, R., & Beard, D. (2018). The nexus of fun and nutrition: Recreational fishing is also about food. *Fish and Fisheries*, *19*(2), 201–224. DOI: 10.1111/faf.12246

Corbett, J. J., & Winebrake, J. J. (2008). Emissions tradeoffs among alternative marine fuels: Total fuel cycle analysis of residual oil, marine gas oil, and marine diesel oil. *Journal of the Air & Waste Management Association*, *58*(4), 538–542. DOI: 10.3155/1047-3289.58.4.538

Culhane, F., Austen, M. C., Ashley, M., Javier, J., Kuit, S. H., Hung, N. P., Tran, H. D., Praptiwi, R. A., Sainal, S., Justine, E., Wulandari, P., Broszeit, S., Jontila, J. B., Johari, S., Creencia, L., Then, A. Y.-H., Gajardo, L. J., Maharja, C., Goh, H. C., & Langmead, O. (2024). Assessing impact risk to tropical marine ecosystems from human activities with a Southeast Asian example. *Journal of Applied Ecology*, *61*(12), 2897–2911. DOI: 10.1111/1365-2664.14812

Cusack, C., Sethi, S. A., Rice, A. N., Warren, J. D., Fujita, R., Ingles, J., Flores, J., Garchitorena, E., & Mesa, S. V. (2021). Marine ecotourism for small pelagics as a source of alternative income generating activities to fisheries in a tropical community. *Biological Conservation*, *261*, 109242. DOI: 10.1016/j.biocon.2021.109242

Dal, N., & Baysan, S. (2007). Coastal Land Use and Attitudes of Local People on Spatial Impacts of Tourism in Kuşadası. *Aegean Geography Journal*, *16*(1-2), 69–85.

Danlu, L. (2023). A Study on the Talents Cultivation Model of Vocational Tourism English Students Based on the Integration of Industry and Education from the perspective of Stakeholders. *Frontiers in Educational Research*, *6*(16).

Davis, D., & Tisdell, C. (1996). Economic management of recreational scuba diving and the environment. *Journal of Environmental Management*, *48*(3), 229–248. DOI: 10.1006/jema.1996.0075

De la Cruz-Modino, R., & Cosentino, M. (2022). Conservation Hub: The Added Value of the Whale-Watching Industry. *Sustainability (Basel)*, *14*(20), 13471. DOI: 10.3390/su142013471

De Melo Rodríguez, G., Martin-Alcalde, E., Murcia-González, J. C., & Saurí, S. (2017). Evaluating air emission inventories and indicators from cruise vessels at ports. *WMU Journal of Maritime Affairs*, *16*(3), 405–420. DOI: 10.1007/s13437-016-0122-8

Dimmock, K., & Musa, G. (2015). Scuba Diving Tourism System: A framework for collaborative management and sustainability. *Marine Policy*, *54*, 52–58. DOI: 10.1016/j.marpol.2014.12.008

Dimopoulos, D., Queiros, D., & Zyl, C. V. (2025). Navigating the ecological decline of coral reef ecosystems and its impact on dive tourism in the East African Marine Ecoregion. *Tourism in Marine Environments*, *20*(2), 149–168. DOI: 10.3727/216901925X17460192677311

Dogan, H. Z. (1989). Forms of adjustment: Sociocultural impacts of tourism. *Annals of Tourism Research*, *16*(2), 216–236. DOI: 10.1016/0160-7383(89)90069-8

Dong, Y. (2025). The impact of tourists on the marine environment: A review and managerial implications. *Regional Environmental Change*, *25*(2), 49. DOI: 10.1007/s10113-025-02385-x

Doxey, G. V. (1975). A causation theory of visitor–resident irritants: Methodology and research inferences. In *Proceedings of the Travel Research Association Sixth Annual Conference*, San Diego, CA, 195–198.

Dreizis, Y. (2020). Water quality management in the coastal zone of the sea. In *E3S Web of Conferences* (Vol. 215, p. 04003). EDP Sciences.

Faulkner, B., & Tideswell, C. (1997). A framework for monitoring community impacts of tourism. *Journal of Sustainable Tourism*, *5*(1), 3–28. DOI: 10.1080/09669589708667273

Fernández, P. R., & Rivera, A. (2024). SCUBA diver behavior, diver carrying capacity and its implications for sustainable reef management. *Marine Policy*, *165*, 106199. DOI: 10.1016/j.marpol.2024.106199

Garcés-Ordóñez, O., Díaz, L. F. E., Cardoso, R. P., & Muniz, M. C. (2020). The impact of tourism on marine litter pollution on Santa Marta beaches, Colombian Caribbean. *Marine Pollution Bulletin*, *160*, 111558. DOI: 10.1016/j.marpolbul.2020.111558

Garrod, B., & Wilson, J. C. (2004). Nature on the edge? Marine ecotourism in peripheral coastal areas. *Journal of Sustainable Tourism*, *12*(2), 95–120. DOI: 10.1080/09669580408667227

Gedik, S., & Mugan-Ertugral, S. (2019). The effects of marine tourism on water pollution. *Fresenius Environmental Bulletin*, *28*(2), 863–866.

GhulamRabbany, M., Afrin, S., Rahman, A., Islam, F., & Hoque, F. (2013). Environmental effects of tourism. American Journal of Environment. *Energy and Power Research*, *1*(7), 117–130.

Gilliland, T. E., Sanchirico, J. N., & Taylor, J. E. (2020). Market-driven bioeconomic general equilibrium impacts of tourism on resource-dependent local economies: A case from the western Philippines. *Journal of Environmental Management, 271*, 110968. DOI: 10.1016/j.jenvman.2020.110968

Ginting, A., & Jayanti, J. (2024). *Influence of communication and attitudes of foreign tourists towards local communities. Vernacular: Linguistics.* Literature, Communication and Culture Journal., DOI: 10.35447/vernacular.v3i2.827

Gissi, E., Manea, E., Mazaris, A. D., Fraschetti, S., Almpanidou, V., Bevilacqua, S., Coll, M., Guarnieri, G., Lloret-Lloret, E., Pascual, M., Petza, D., Rilov, G., Schonwald, M., Stelzenmüller, V., & Katsanevakis, S. (2021). A review of the combined effects of climate change and other local human stressors on the marine environment. *The Science of the Total Environment, 755*, 142564. DOI: 10.1016/j.scitotenv.2020.142564

Gladstone, W., Curley, B., & Shokri, M. R. (2013). Environmental impacts of tourism in the Gulf and the Red Sea. *Marine Pollution Bulletin, 72*(2), 375–388. DOI: 10.1016/j.marpolbul.2012.09.017

Gobler, C. J. (2020). Climate change and harmful algal blooms: Insights and perspective. *Harmful Algae, 91*, 101731. DOI: 10.1016/j.hal.2019.101731

Gomes, C. S., Fernandes, J. V., Fernandes, F. V., & Silva, J. B. (2021). Salt mineral water and thalassotherapy. In *Minerals latu sensu and Human Health: Benefits, Toxicity and Pathologies* (pp. 631–656). Springer International Publishing. DOI: 10.1007/978-3-030-65706-2_16

Gooden, A., Clarke, T. M., Meyer, L., & Huveneers, C. (2024). Wildlife tourism has little energetic impact on the world's largest predatory shark. *Animal Behaviour, 207*, 247–265. DOI: 10.1016/j.anbehav.2023.10.004

Görmüş, K. S., Kutoğlu, Ş. H., Şeker, D. Z., Özölçer, İ. H., Oruç, M., & Aksoy, B. (2014). Temporal analysis of coastal erosion in Turkey: A case study Karasu coastal region. *Journal of Coastal Conservation, 18*(4), 399–414. DOI: 10.1007/s11852-014-0325-0

Gössling, S. (2002). Global environmental consequences of tourism. *Global Environmental Change, 12*(4), 283–302. DOI: 10.1016/S0959-3780(02)00044-4

Gössling, S., Hall, C. M., & Scott, D. (2018). Coastal and ocean tourism. *Handbook on marine environment protection: Science, impacts and sustainable management*, 773-790.

Greenwood, D. J. (1989). Culture by the pound: An anthropological perspective on tourism as cultural commodification. In Smith, V. (Ed.), *Hosts and Guests: The Anthropology of Tourism* (pp. 171–185). University of Pennsylvania Press.

Guo, S., Feng, X., & Zhou, B. (2024). Trends and perspectives of marine sports tourism: A bibliometric analysis and systematic review. *Water* (20734441), 16(19).

Hall, C. M. (2001). Trends in ocean and coastal tourism: The end of the last frontier? *Ocean and Coastal Management, 44*(9-10), 601–618. DOI: 10.1016/S0964-5691(01)00071-0

Hayati, Y., Adrianto, L., Krisanti, M., Pranowo, W. S., & Kurniawan, F. (2020). Magnitudes and tourist perception of marine debris on small tourism island: Assessment of Tidung Island, Jakarta, Indonesia. *Marine Pollution Bulletin, 158*, 111393. DOI: 10.1016/j.marpolbul.2020.111393

Hynes, S., Cawley, M., Deely, J., & Norton, D. (2024). Alternative approaches to measuring the value of tourism in marine and coastal areas in ocean economy accounting. *Marine Policy, 168*, 106299. DOI: 10.1016/j.marpol.2024.106299

Ioras, F. (2025). Sustainable Management of Mediterranean Superyacht Marinas: A Comparative Assessment of Environmental Practices and Policy Implications. *Sustainability (Basel), 17*(4), 1377. DOI: 10.3390/su17041377

Ioras, F., & Bandara, I. (2025). Artificial Intelligence and Sustainable Practices in Coastal Marinas: A Comparative Study of Monaco and Ibiza. *Sustainability (Basel), 17*(16), 7404. DOI: 10.3390/su17167404

Jambeck, J. R., Geyer, R., Wilcox, C., Siegler, T. R., Perryman, M., Andrady, A., ... & Law, K. L. (2015). Plastic waste inputs from land into the ocean. science, 347(6223), 768-771.

Jennings, G. (2007). Water-based tourism, sport, leisure, and recreation experiences. In *Water-based tourism, sport, leisure, and recreation experiences* (pp. 1–20). Routledge. DOI: 10.4324/9780080468310-7

Ji, X., & Ding, X. (2024). Analysis on the relationship between coastal tourism and marine pollution: An empirical analysis of China's 11 coastal regions. *Frontiers in Marine Science, 11*, 1471467. DOI: 10.3389/fmars.2024.1471467

Jonas, L. C. (2024). Emerging Opportunities in Blue Tourism. *Tourism in Marine Environments, 19*(1), 59–67. DOI: 10.3727/216901924X17117021040174

Kabil, M., Ali, M. A., Marzouk, A., & Dávid, L. D. (2024). Gender perspectives in tourism studies: A comparative bibliometric analysis in the MENA region. *Tourism Planning & Development, 21*(3), 276–298. DOI: 10.1080/21568316.2022.2050419

Kaczynski, W. (2011). The future of blue economy: Lessons for European Union. *Foundations of Management, 3*(1), 21–32. DOI: 10.2478/v10238-012-0033-8

Kasim, M., Oetama, D., Bahar, A., Alelo, M., Wardani, K. K., & Susandari, H. (2021). The role of local communities in developing potential areas for marine tourism. []. IOP Publishing.]. *IOP Conference Series. Earth and Environmental Science, 800*(1), 012055. DOI: 10.1088/1755-1315/800/1/012055

Kerr, J. T., & Ostrovsky, M. (2003). From space to species: Ecological applications for remote sensing. *Trends in Ecology & Evolution, 18*(6), 299–305. DOI: 10.1016/S0169-5347(03)00071-5

Kester, J. G. (2003). Cruise tourism. *Tourism Economics, 9*(3), 337–350. DOI: 10.1177/135481660300900307

Kim, A., Scott, C. P., & Swartz, W. (2024). Local perspectives on marine ecotourism development in a water-insecure island region: The case of Bocas del Toro, Panama. *Frontiers in Marine Science, 11*, 1377053. DOI: 10.3389/fmars.2024.1377053

Kim, M., & Boo, C. S. (2015). An exploratory study on visit intention of destination in marine health tourism. *Susan Haeyang Gyoyuk Yeon-Gu, 27*(1), 230–242. DOI: 10.13000/JFMSE.2015.27.1.230

Kizielewicz, J., & Luković, T. (2013). The phenomenon of the marina development to support the European model of economic development. *TransNav. International Journal on Marine Navigation and Safety Od Sea Transportation, 7*(3), 461–466. DOI: 10.12716/1001.07.03.19

Kolesnikova, M. (2020). EU maritime economy and COVID-19. *Sovremennaya Yevropa, 2020*(4), 102–111. DOI: 10.15211/soveurope42020102111

Kothalawala, H. S., & Arachchi, R. S. S. W. (2021). Potential of developing scuba diving tourism in Sri Lanka: Scuba diving operators perspective. *Journal of Tourism Economics and Applied Research, 5*(1), 10–28.

Kovačić, M., & Silveira, L. (2020). Cruise tourism: Implications and impacts on the destinations of Croatia and Portugal. *Pomorstvo, 34*(1), 40–47. DOI: 10.31217/p.34.1.5

Kurniawan, F., Adrianto, L., Bengen, D. G., & Prasetyo, L. B. (2023). Hypothetical effects assessment of tourism on coastal water quality in the Marine Tourism Park of the Gili Matra Islands, Indonesia. *Environment, Development and Sustainability, 25*(8), 7959–7985. DOI: 10.1007/s10668-022-02382-8

Lan, N. T. P., & Thanh, D. T. N. (2023). Maritime security or economic growth: The choice of stakeholders in coastal tourism development in Binh Dai district, Ben Tre province, Vietnam. [). IOP Publishing.]. *IOP Conference Series. Earth and Environmental Science, 1247*(1), 012005. DOI: 10.1088/1755-1315/1247/1/012005

Łapko, A., Strulak-Wójcikiewicz, R., Landowski, M., & Wieczorek, R. (2018). Management of waste collection from yachts and tall ships from the perspective of sustainable water tourism. *Sustainability (Basel), 11*(1), 121. DOI: 10.3390/su11010121

Li, L., Wu, B., & Patwary, A. K. (2022). How marine tourism promote financial development in sustainable economy: New evidences from South Asia and implications to future tourism students. *Environmental Science and Pollution Research International, 29*(1), 1155–1172. DOI: 10.1007/s11356-021-15709-1

Liang, X. C., & Ding, D. S. (2002). Trends of overseas studies of ocean and coastal tourism. *Ziran Ziyuan Xuebao, 17*(6), 783–791.

Liu, W., & Cao, Z. (2018). Positive role of marine tourism on economic stimulus in coastal area. *Journal of Coastal Research*, (83), 217–220.

Lopez-Rivas, J. D., & Cardenas, J. C. (2024). What is the economic value of coastal and marine ecosystem services? A systematic literature review. *Marine Policy, 161*, 106033. DOI: 10.1016/j.marpol.2024.106033

MacCannell, D. (1976). *The Tourist: A New Theory of the Leisure Class*. University of California Press.

Machado, J. T. M., & de Andrés, M. (2023). Implications of offshore wind energy developments in coastal and maritime tourism and recreation areas: An analytical overview. *Environmental Impact Assessment Review, 99*, 106999. DOI: 10.1016/j.eiar.2022.106999

Maione, C. (2021). Quantifying plastics waste accumulations on coastal tourism sites in Zanzibar, Tanzania. *Marine Pollution Bulletin, 168*, 112418. DOI: 10.1016/j.marpolbul.2021.112418

Martínez Vázquez, R. M., Milán García, J., & De Pablo Valenciano, J. (2021). Analysis and Trends of Global Research on Nautical, Maritime and Marine *Tourism. Journal of Marine Science and Engineering, 9*(1), 93. DOI: 10.3390/jmse9010093

Mbaiwa, J. E. (2005). The socio-cultural impacts of tourism development in the Okavango Delta, Botswana. *Journal of Tourism and Cultural Change*, *2*(3), 163–185. DOI: 10.1080/14766820508668662

McKegg, S. T. (1999). *Marine tourism in New Zealand: environmental issues and options* [Unpublished doctoral dissertation] University of Otago, New Zealand

McKercher, B., & Du Cros, H. (2002). *Cultural tourism: The partnership between tourism and cultural heritage management*. Routledge.

Mejjad, N., Laissaoui, A., Fekri, A., & Hammoumi, O. E. (2023). Marine plastic pollution in Morocco: State of the knowledge on origin, occurrence, fate, and management. *Environmental Science and Pollution Research International*, *30*(49), 107371–107389. DOI: 10.1007/s11356-023-26973-8

Merino, F., & Prats, M. A. (2022). Are blue flags a good indicator of the quality of sea water on beaches? An empirical analysis of the Western Mediterranean basin. *Journal of Cleaner Production*, *330*, 129865. DOI: 10.1016/j.jclepro.2021.129865

Middleton, V. T., & Hawkins, R. (1998). *Sustainable tourism: A marketing perspective*. Routledge.

Miller, M. L., & Auyong, J. (1991). Coastal zone tourism: A potent force affecting environment and society. *Marine Policy*, *15*(2), 75–99. DOI: 10.1016/0308-597X(91)90008-Y

Moric, I., Pekovic, S., Janinovic, J., Perovic, Đ., & Griesbeck, M. (2021). Cultural tourism and community engagement: Insight from Montenegro. Business Systems Research: *International journal of the Society for Advancing Innovation and Research in Economy*, 12(1), 164-178.

Musyafah, A., Prasetyo, M., Prananda, R., & Jati, O. (2025). Challenges in Implementing Sustainable Marine Tourism Policy against the Threat of Waste in Coastal Areas. *International Journal of Multidisciplinary Research And Analysis*. ..DOI: 10.47191/ijmra/v8-i01-20

Nababan, Y. N. (2023). Scuba diving as sustainability economic marine ecotourism: Scuba diving as sustainability economic marine ecotourism. *Journal of Oceanography and Aquatic Science*, *1*(2), 28–32. DOI: 10.56855/joane.v1i2.338

Nedelcu, A., Popescu, G., & Cioaca, A. (2015). Current diversity of cruise tourism. *Journal of Tourism Challenges and Trends*, *8*(1), 27–42.

Nguyen, C., Schinckus, C., Su, T., & Chong, F. (2020). The Influence of Tourism on Income Inequality. *Journal of Travel Research, 60*(7), 1426–1444. DOI: 10.1177/0047287520954538

Nguyen, D., Nguyen, M. K., Pham, M. T., Anh, N. T., Van Tri, D., Nguyen, H. L., & Nguyen, D. D. (2025). Investigating macro marine litter and beach cleanliness along Southern Vietnam beaches. *Marine Pollution Bulletin, 212*, 117566. DOI: 10.1016/j.marpolbul.2025.117566

OECD. (2016). *Environmental impacts of cruise tourism.* OECD Publishing.

Onofri, L., & Nunes, P. A. (2013). Beach 'lovers' and 'greens': A worldwide empirical analysis of coastal tourism. *Ecological Economics, 88*, 49–56. DOI: 10.1016/j.ecolecon.2013.01.003

Orams, M. (2002). *Marine tourism: development, impacts and management.* Routledge. DOI: 10.4324/9780203197110

Otrachshenko, V., & Bosello, F. (2017). Fishing for answers? Impacts of marine ecosystem quality on coastal tourism demand. *Tourism Economics, 23*(5), 963–980. DOI: 10.1177/1354816616656422

Papathanassis, A. (2020). The growth and development of the cruise sector: A perspective article. *Tourism Review, 75*(1), 130–135. DOI: 10.1108/TR-02-2019-0037

Papathanassis, A. (2023). A decade of 'blue tourism'sustainability research: Exploring the impact of cruise tourism on coastal areas. *Cambridge Prisms: Coastal Futures, 1*, e12. DOI: 10.1017/cft.2023.2

Pawson, M. G., Glenn, H., & Padda, G. (2008). The definition of marine recreational fishing in Europe. *Marine Policy, 32*(3), 339–350. DOI: 10.1016/j.marpol.2007.07.001

Pessoa, R. M. C., Jiménez, J. A., da Costa, R. M., & Pereira, L. C. C. (2019). Federal conservation units in the Brazilian amazon coastal zone: An adequate approach to control recreational activities? *Ocean and Coastal Management, 178*, 104856. DOI: 10.1016/j.ocecoaman.2019.104856

Pham, T. (2020). Tourism in marine protected areas: Can it be considered as an alternative livelihood for local communities? *Marine Policy, 115*, 103891. DOI: 10.1016/j.marpol.2020.103891

Picken, F. (2025). Tourism and the blue economy. *Tourism Geographies, 27*(3-4), 538–546. DOI: 10.1080/14616688.2023.2291821

Prihadi, D. J., Dhahiyat, A. P., Kristiadhi, F., & Pamungkas, W. (2024). Sustainable tourism in 2024: The Role of tourists and industry in Preserving marine nature. *MSJ: Majority Science Journal, 2*(4), 112–118. DOI: 10.61942/msj.v2i4.252

Radlińska, K., & Gardziejewska, B. (2022). The seasonal labor hoarding in tourist enterprises—Choice or necessity? *Sustainability (Basel), 14*(12), 6995. Advance online publication. DOI: 10.3390/su14126995

Rangel-Buitrago, N., Williams, A. T., Ergin, A., Anfuso, G., Micallef, A., & Pranzini, E. (2019). Coastal scenery: An introduction. *Coastal Scenery: Evaluation and Management*, 1-16.

Razak, M. A., Ghasemlouia, R., & Aksel, M. (2023). Tourism impact on coastal erosion: A case of Alanya. *International Journal of Environment and Geoinformatics, 10*(4), 105–116. DOI: 10.30897/ijegeo.1287569

Rhoden, S., & Kaaristo, M. (2020). Liquidness: Conceptualising water within boating tourism. *Annals of Tourism Research, 81*, 102854. DOI: 10.1016/j.annals.2019.102854

Roberts, C. M., O'Leary, B. C., McCauley, D. J., Cury, P. M., Duarte, C. M., Lubchenco, J., Pauly, D., Sáenz-Arroyo, A., Sumaila, U. R., Wilson, R. W., Worm, B., & Castilla, J. C. (2017). Marine reserves can mitigate and promote adaptation to climate change. *Proceedings of the National Academy of Sciences of the United States of America, 114*(24), 6167–6175. DOI: 10.1073/pnas.1701262114

Saveriades, A. (2000). Establishing the social tourism carrying capacity for the tourist resorts of the east coast of the Republic of Cyprus. *Tourism Management, 21*(2), 147–156. DOI: 10.1016/S0261-5177(99)00044-8

Schneider, F., Parsons, S., Clift, S., Stolte, A., & McManus, M. C. (2018). Collected marine litter—A growing waste challenge. *Marine Pollution Bulletin, 128*, 162–174. DOI: 10.1016/j.marpolbul.2018.01.011

Schuhmann, P., Skeete, R., Waite, R., Bangwayo-Skeete, P., Casey, J., Oxenford, H. A., & Gill, D. A. (2019). Coastal and marine quality and tourists' stated intention to return to Barbados. *Water (Basel), 11*(6), 1265. DOI: 10.3390/w11061265

Sevinç, F., & Güzel, T. (2017). Sustainable yacht tourism practices. *Management & Marketing Journal,* 15(1).

Sharpley, R. (2014). Host perceptions of tourism: A review of the research. *Tourism Management, 42*, 37–49. DOI: 10.1016/j.tourman.2013.10.007

Simonsen, M., Gössling, S., & Walnum, H. J. (2019). Cruise ship emissions in Norwegian waters: A geographical analysis. *Journal of Transport Geography, 78*, 87–97. DOI: 10.1016/j.jtrangeo.2019.05.014

Smith, M. K. (2015). *Issues in cultural tourism studies*. Routledge. DOI: 10.4324/9781315767697

Spalding, M., Burke, L., Wood, S. A., Ashpole, J., Hutchison, J., & Zu Ermgassen, P. (2017). Mapping the global value and distribution of coral reef tourism. *Marine Policy, 82*, 104–113. DOI: 10.1016/j.marpol.2017.05.014

Staffieri, E., de Lucia, G. A., Camedda, A., Poeta, G., & Battisti, C. (2019). Pressure and impact of anthropogenic litter on marine and estuarine reptiles: An updated "blacklist" highlighting gaps of evidence. *Environmental Science and Pollution Research International, 26*(2), 1238–1249. DOI: 10.1007/s11356-018-3616-4

Stepanets, I., & Hryniuk, D. (2021). *Economic and geographical aspects of the development of yachting as a promising sports tourism in the world*. Black Sea Economic Studies., DOI: 10.32843/bses.68-2

Stylidis, D. (2020). Exploring resident–Tourist interaction and its impact on tourists' destination image. *Journal of Travel Research, 61*(1), 186–201. DOI: 10.1177/0047287520969861

Sukran, M. A., Kurniawan, T., Basri, H., & Furqan, A. (2025). Exploring marine tourism on sustainable development segment in Indonesia: SPAR-4-SLR. *Geo Journal of Tourism and Geosites, 58*(1), 243–256. DOI: 10.30892/gtg.58121-1406

Swarbrooke, J. (2020). *The Impact of Tourism on the Marine Environment*. Goodfellow Publishers Ltd.

Telfer, D. J., & Sharpley, R. (2015). *Tourism and development in the developing world*. Routledge. DOI: 10.4324/9781315686196

Thi, H. N., Thi, T. N., Trong, T. V., Duc, T. N., & Nghi, T. N. (2024). Sustainable tourism governance: A study of the impact of culture. *J. Gov. Regul, 13*(2).

Timothy, D. J. (2006). Empowerment and stakeholder participation in tourism destination communities. In *Tourism, power and space* (pp. 213–230). Routledge.

Timothy, D. J., & Boyd, S. W. (2003). *Heritage Tourism*. Pearson Education.

Torres, R., & Momsen, J. H. (2004). Challenges and potential for linking tourism and agriculture to achieve pro-poor tourism objectives. *Progress in Development Studies, 4*(4), 294–318. DOI: 10.1191/1464993404ps092oa

Tosun, C. (2000). Limits to community participation in the tourism development process in developing countries. *Tourism Management, 21*(6), 613–633. DOI: 10.1016/S0261-5177(00)00009-1

Tsai, F. M., Bui, T. D., Tseng, M. L., Lim, M. K., & Tan, R. R. (2021). Sustainable solid-waste management in coastal and marine tourism cities in Vietnam: A hierarchical-level approach. *Resources, Conservation and Recycling, 168*, 105266. DOI: 10.1016/j.resconrec.2020.105266

UNEP. (2006). *Marine and coastal ecosystems and human wellbeing: A synthesis report based on the findings of the Millennium Ecosystem Assessment.* UNEP.

UNEP. (2009). *Sustainable Coastal Tourism: an integrated planning and management approach.* UNEP.

UNESCO. (2003). *Convention for the Safeguarding of the Intangible Cultural Heritage.* United Nations Educational, Scientific and Cultural Organization.

Unhasuta, S., Sasaki, N., & Kim, S. M. (2021). Impacts of tourism development on coastal communities in Cha-am Beach, the Gulf of Thailand, through analysis of local perceptions. *Sustainability (Basel), 13*(8), 4423. DOI: 10.3390/su13084423

Van Putten, I., Metcalf, S., Frusher, S., Marshall, N., & Tull, M. (2014). Fishing for the impacts of climate change in the marine sector: A case study. *International Journal of Climate Change Strategies and Management, 6*(4), 421–441. DOI: 10.1108/IJCCSM-01-2013-0002

Velando, A., & Munilla, I. (2011). Disturbance to a foraging seabird by sea-based tourism: Implications for reserve management in marine protected areas. *Biological Conservation, 144*(3), 1167–1174. DOI: 10.1016/j.biocon.2011.01.004

Vetrimurugan, E., Shruti, V. C., Jonathan, M. P., Roy, P. D., Kunene, N. W., & Villegas, L. E. C. (2017). Metal concentration in the tourist beaches of South Durban: An industrial hub of South Africa. *Marine Pollution Bulletin, 117*(1-2), 538–546. DOI: 10.1016/j.marpolbul.2017.02.036

Wang, Y., & Zhang, H. (2020). Influence of Cultural Identity on Tourists' Authenticity Perception and Destination Loyalty in Intangible Cultural Heritage Tourism. *Sustainability, 12*(16), 6344. DOI: 10.3390/su12166344

Weatherdon, L. V., Magnan, A. K., Rogers, A. D., Sumaila, U. R., & Cheung, W. W. (2016). Observed and projected impacts of climate change on marine fisheries, aquaculture, coastal tourism, and human health: An update. *Frontiers in Marine Science, 3*, 48. DOI: 10.3389/fmars.2016.00048

Williams, A. M., & Hall, C. M. (2000). Tourism and migration: New relationships between production and consumption. *Tourism Geographies*, *2*(1), 5–27. DOI: 10.1080/146166800363420

Wilson, J. C., & Garrod, B. (2003). Introduction. In Garrod, B., & Wilson, J. C. (Eds.), *Marine ecotourism: issues and experiences* (Vol. 7). Channel View Publications. DOI: 10.21832/9781873150436-004

World Tourism Organization (2024). *International Tourism Highlights, 2024 Edition*, November 2024, UN Tourism, Madrid, DOI:.DOI: 10.18111/9789284425808

Yang, J., Ryan, C., & Zhang, L. (2013). Social conflict in communities impacted by tourism. *Tourism Management*, *35*, 82–93. DOI: 10.1016/j.tourman.2012.06.002

Zentner, Y., Margarit, N., Ortega, J., Casals, D., Medrano, A., Pagès-Escolà, M., & Linares, C. (2023). Marine protected areas in a changing ocean: Adaptive management can mitigate the synergistic effects of local and climate change impacts. *Biological Conservation*, *282*, 110048. DOI: 10.1016/j.biocon.2023.110048

Zheng, Y., & Liu, D. (2021). Research on marine pollution problems and solutions in China from the perspective of marine tourism. *Journal of Marine Science*, *3*(1). Advance online publication. DOI: 10.30564/jms.v3i1.2599

KEY TERMS AND DEFINITIONS

Ecological Degradation: Loss of quality, decrease in functionality and unsustainable changes in the structure, function, species composition and ecosystem services of the natural environment due to human activities.

Socio-cultural Impacts: It refers to the positive or negative effects of tourism activities on the social structure, cultural values, lifestyles and social relations of local communities.

Stakeholder Participation: It is the systematic determination of the knowledge, values and needs of individuals and groups that affect or are affected by decision-making processes and their inclusion in the process.

Sustainability: It is a concept that aims to meet the needs of people today without endangering the needs of future generations, while maintaining a balance in environmental, economic and social (or human) dimensions, and to ensure the long-term protection of natural resources and life support systems.

Chapter 7
Climate Change and Willingness to Pay for Beach Conservation:
Insights From Costa del Sol (Spain)

Laura Banks
University of Malaga, Spain

Alfonso Expósito
https://orcid.org/0000-0002-9248-4879
University of Malaga, Spain

ABSTRACT

Climate change (CC) and the increasing occurrence of extreme weather events present significant threats to coastal regions, particularly those reliant on tourism. This chapter analyses the case of the Mediterranean Costa del Sol in southern Spain, a destination where tourism revenues and attractiveness are closely tied to the environmental quality of its coastline. The study investigates beach users' willingness to pay (WTP) to support conservation and adaptation measures aimed at countering CC impacts. Findings show that visitors are generally willing to contribute financially to beach conservation, though this disposition is shaped by socio-economic and attitudinal factors, including income, level of education, and environmental concern. For coastal municipalities, the central challenge is to transform this willingness into practical financing mechanisms that are socially accepted, transparent, and equitable. This study provides critical evidence for designing inclusive coastal policies and awareness strategies to strengthen the climate resilience of beach destinations.

DOI: 10.4018/979-8-3373-1777-9.ch007

INTRODUCTION

Climate change (CC) represents a growing challenge for tourism activities and marine ecosystems both in Spain and globally, as climatic conditions are undergoing rapid and profound transformations. In the case of the Mediterranean Costa del Sol, located in southern Spain and recognized as one of Europe's leading destinations for sun-and-beach tourism, these impacts are particularly pronounced. Rising sea levels, accelerated coastal erosion, and the increasing frequency and intensity of extreme weather events, such as heatwaves and storms, are placing considerable pressure on the region's environmental resources and, consequently, on its long-term attractiveness as a tourist destination. These challenges threaten coastal infrastructure, economic activity, and beach quality, ultimately reducing the region's appeal as tourist destination. Rising temperatures are already affecting the Costa del Sol, potentially making summers too hot for some visitors. Warmer waters can also lead to declining water quality and increased beach safety risks due to the proliferation of harmful algae (Mycoo & Gobin, 2013). Changes in precipitation and rising nighttime temperatures further impact tourism dynamics, potentially altering visitor patterns (Do Ó, A., & Seiz R., 2021). According to IPCC projections, temperatures in the region could rise by 1.5°C between 2030 and 2052 if global warming continues at its current pace (IPCC, 2019). Additionally, the average temperature of the Mediterranean Sea could increase between 1.5°C and 4.5°C, depending on greenhouse gas emission scenarios (Pérez-Juan et al., 2022).

Coastal erosion and sea level rise constitute good examples of CC impacts on coastal destinations (Vitousek et al., 2017). Coastal erosion refers to the progressive loss and displacement of shoreline materials, such as sand, gravel, and rocks, due to natural processes like wave action and ocean currents. Human activities, such as coastal construction, and CC exacerbate this phenomenon (Ramdas & Mohamed, 2014). According to Spain's Coastal Regulations, beach regression rates range from 1.5 to 4.5 meters per year, with some areas experiencing retreat exceeding five meters (Rejón, 2024). These climate-induced changes have significant socioeconomic consequences, including reduced tourism attraction, declining revenues for coastal businesses, and increased costs associated with beach cleanup and recovery efforts. Additionally, the degradation of beach ecosystems affects biodiversity, leading to the disappearance of native species and the proliferation of non-native organisms that disrupt ecological balance. Rising sea levels and extreme weather events further stress coastal infrastructure, requiring expensive adaptation measures to minimize damages.

This study aims to assess the willingness to pay (WTP) of beach users to contribute to financing conservation measures in response to the impacts of CC and extreme weather events. Additionally, our assessment aims to explore the factors (e.g., socio-economic, attitudinal, CC awareness) that determine respondents' WTP. The case study of the coastal town of Fuengirola in the Spanish Costa del Sol has been selected to fulfill these research objectives. The study contributes to Sustainable Development Goal (SDG) 13 "Climate Action" by exploring sustainable financing mechanisms for coastal protection. Additionally, it underscores beaches' economic and environmental significance as crucial assets for local communities. The results reveal significant correlations between WTP and various socioeconomic and environmental factors. Interestingly, higher-income, environmentally concerned, and frequent beach visitors are more willing to pay for conservation measures. Foreign tourists are willing to pay more than national visitors. Regarding the respondent's gender, women demonstrate slightly higher WTP than men. Visitors' educational background also plays a role; higher-educated respondents exhibited higher WTP, and those who expressed strong concern for CC were more inclined to contribute financially. These findings highlight the importance of tailoring coastal management policies to different demographic groups, ensuring equitable contributions to conservation efforts.

This chapter is organized as follows. The next section provides a brief review of recent comparable studies conducted in other coastal locations. This is followed by the presentation of the case study of Fuengirola (Spain), together with a detailed description of the data, materials, and methodological approach employed in the analysis. The subsequent section outlines the main empirical findings, which are then discussed in light of their policy implications, with particular attention to recommendations for coastal management and climate adaptation. The chapter concludes with a set of final remarks that synthesize the key contributions and highlight avenues for future research.

LITERATURE REVIEW

The economic valuation of coastal resources, particularly through contingent valuation (CV) methods, has become a widely applied tool to assess visitors' WTP for environmental improvements and conservation actions. CV provides a framework to capture the non-market value of environmental goods by eliciting preferences directly from users and visitors (Mitchell & Carson, 1989; Hanley & Barbier, 2009). In the context of CC and its impacts on coastal areas, CV has been employed to understand how beach users perceive risks such as erosion, pollution, and ecosystem degradation, and how much they are willing to contribute financially to mitigation and adaptation strategies (Nesha Dushani et al., 2023; Onofrio et al., 2025).

Several case studies highlight the diversity of contexts, methods, and outcomes when applying CV to coastal environments. For example, Alves et al. (2015) conducted a survey with 765 users across three beaches in Cádiz (Spain), focusing on perceptions of erosion and preferences for improved beach management. While respondents exhibited high awareness of coastal erosion, only 12.9% expressed a positive WTP, attributing their refusal primarily to economic constraints, perceptions of high taxation, and the belief that conservation should be financed through existing public mechanisms. Similar results were highlighted by Schuhmann et al. (2024) using data from a CV model and a discrete choice experiment carried out in Grenada (Caribbean Sea), where more than half of respondents were reluctant to donate for conservation purposes due to a wide variety of reasons. This outcome illustrates how socio-economic context and attitudes towards governance can limit the financial feasibility of user-based mechanisms.

By contrast, Kontogianni et al. (2014) studied two beaches on Lesvos Island (Greece) affected by beach rock formation, surveying 106 European tourists. They reported a much higher level of acceptance, with 47.2% of respondents willing to pay, and estimated average household contributions ranging from €13.2 to €16.4 annually. Importantly, they identified explanatory factors, such as participation in water sports, marital status, and employment conditions, that positively influenced visitors' WTP. These findings underscore the role of visitor characteristics and recreational preferences in shaping economic valuation outcomes.

Another perspective is provided by Birdir et al. (2013) in Mersin (Turkey), where 402 beach users were surveyed to assess their WTP for improvements in cleanliness, provision of social activities, and maintenance of landscape quality. Average WTP values per visit ranged from €1.77 to €2.33 depending on the beach, and over 92% of respondents declared a positive WTP. Younger visitors and more frequent beach users were particularly willing to contribute. This case contrasts with the study of Alves et al. (2015), illustrating that WTP can be substantially higher in contexts where tangible improvements, such as enhanced services and amenities, are directly visible and valued by users.

Studies, such as those of Enriquez-Acevedo et al. (2018), focused on the case of three beaches in the Colombian Caribbean (Puerto Velero, Caño Dulce and Salgar), and Choi et al. (2021) addressing the case of Haeundae beach in Korea, have pointed out that a majority of visitors usually expresses a positive WTP, suggesting a strong commitment to preserving ecosystem services. Environmental awareness, perceived vulnerability to ecosystem loss, and satisfaction with the beach were among the strongest predictors of WTP.

A broader national-scale analysis was undertaken by Rodella et al. (2019), who surveyed more than 5,000 users across 41 beaches in 11 Italian coastal regions. Their results indicated a national average WTP of €14.84 per year, with 58% of respondents willing to contribute. Age and gender emerged as significant predictors, with women and older respondents reporting higher WTP values. This large-scale study is particularly relevant as it demonstrates that, despite site-specific differences, CV can provide robust evidence to support policy decisions at regional or national scales.

Finally, in East Asia, Liu et al. (2021) studied 259 visitors to Qingdao beaches (China), focusing on how environmental perceptions and satisfaction influenced WTP. Their findings revealed that visitor satisfaction was positively correlated with WTP and that perceived quality dimensions (particularly intangible aspects such as natural beauty) were the strongest predictors of higher contributions. These results reinforce the idea that perceived environmental quality and visitor experience are central determinants of economic valuation outcomes.

Taken together, these studies reveal several important insights. First, WTP estimates for beach conservation vary widely depending on local socio-economic conditions, beach characteristics, and the type of conservation or service improvement proposed. While studies such as Alves et al. (2015) highlight limitations in contexts where users view conservation as a government responsibility, other studies (Birdir et al., 2013; Enriquez-Acevedo et al., 2018) demonstrate broad support where conservation benefits are perceived as direct, tangible, and immediate. Second, socio-demographic variables such as age, income, nationality, and recreational habits consistently emerge as significant determinants of WTP, underlining the need for tailored policy instruments. Finally, these cases underscore the usefulness of CV method as a policy-relevant tool. By quantifying user preferences and financial commitments, CV provides empirical evidence to guide the design of sustainable financing mechanisms for coastal adaptation and conservation, a need that is especially urgent under accelerating CC pressures (Hanley & Barbier, 2009; Freeman et al., 2014). This study aims to contribute to the scarce literature on the assessment of beach users' WTP (and its determinants) to contribute to beach conservation in a context of increasing impacts of CC on coastal regions. With this aim, this study analyses the case study of Fuengirola, a touristic coastal town on the Spanish Costa del Sol.

CASE STUDY

Fuengirola constitutes an emblematic tourist destination situated on the Mediterranean Costa del Sol, in southern Spain. The Costa del Sol region is highly dependent on tourism and its associated economic sectors, including commerce, hospitality, and real estate. This strong reliance renders the area particularly vulnerable to the impacts of CC, as environmental alterations can directly affect both the physical assets (e.g., beaches, coastline, infrastructure) and the socio-economic foundations of the local economy.

The town of Fuengirola has undergone a profound transformation in the last decades. From its origins as a modest fishing village in the first half of the twentieth century, the city has evolved into an international hub of tourism. This trajectory has been shaped by a combination of geographical, economic, and cultural factors. Its privileged position on the Mediterranean coastline, the availability of extensive sandy beaches, a temperate climate, and a wide array of leisure and cultural services have together consolidated its attractiveness as a global tourism destination. The town is located in the central part of the Costa del Sol, alongside other renowned tourist municipalities such as Benalmádena, Mijas, and Marbella. Despite covering an area of only about 10 km^2, Fuengirola is one of the Spanish cities with the highest population densities, surpassing 8,000 inhabitants per km^2. The municipality has 8 km of beaches fully integrated into the urban landscape. These beaches have received notable recognitions, including the Q for Quality certification and multiple Blue Flag awards, conferred by the European Union in recognition of high standards of maintenance, water quality, and beach services (Fuengirola City Council, 2023).

The coastline of Fuengirola represents one of its most important natural resources, functioning not only as a central attraction for international visitors but also as a recreational and social space for residents. The beaches are equipped with diverse services such as showers, sanitary facilities, beach bars, and designated areas to ensure accessibility for people with reduced mobility (Fuengirola City Council, 2024). Furthermore, the coastal strip combines highly urbanized beaches with heavy tourist influx and other areas that, though less developed, retain considerable ecological and scenic value.

In recognition of its efforts toward sustainability and integrated coastal management, Fuengirola obtained in March 2023 the “S for Sustainable Tourism” certification across all its beaches, awarded by the Spanish Institute for Tourism Quality (ICTE). This distinction reflects the municipality’s commitment to responsible coastal governance and its focus on ensuring high standards in accessibility, environmental quality, and public services (Fuengirola City Council, 2023). Such achievements position Fuengirola not only as a consolidated tourist hub but also as a case study for exploring the interplay between coastal tourism development, environmental sustainability, and CC adaptation in Mediterranean destinations.

As previously noted, the Spanish Mediterranean coast -including the Costa del Sol- is experiencing severe impacts from CC, manifested through increasingly frequent and intense extreme weather events such as storms, flooding, and coastal erosion (Laghrissi, 2024). During the 2024–2025 period, several such events produced considerable damage along the Fuengirola coastline. These disturbances left the beaches covered with dense accumulations of organic debris, a visible sign of erosion processes and the disruption of marine ecosystems. The coastline itself has retreated, and the morphology of the beaches has undergone substantial alteration. The accumulation of algae and other plant matter, transported by strong winds, wave action, and sea currents, reflects both the intensity of the meteorological events and the vulnerability of the coastal ecosystems. In addition to environmental impacts, the storms left parasols, loungers, and other beach infrastructure abandoned and entangled within the debris, further evidencing the destructive force of the waves. Beyond aesthetic degradation, the deposition of organic matter disrupts natural sedimentation processes and has potential negative implications for local biodiversity and ecosystem functioning.

In response, the municipality of Fuengirola has taken important steps to reinforce its resilience to CC in the last years. Local authorities have adopted a series of municipal climate action plans, sustainability initiatives, and beach regeneration projects supported by local, national, and European funding sources. These initiatives demonstrate a clear institutional commitment to safeguarding the coastline for future generations while also maintaining the city's competitiveness as a premier Mediterranean tourist destination. Notable measures include:

- Beach maintenance and cleaning: Daily cleaning operations and regular sand regeneration activities to counteract erosion and preserve beach quality.
- Marine ecosystem protection: Sustainable development programmes designed to conserve marine habitats and promote eco-friendly forms of tourism.
- Sustainable infrastructure development: Efforts to preserve and upgrade coastal promenades and related infrastructure, integrating innovations such as solar-powered watchtowers and projects that enhance accessibility and sustainability.
- Community awareness and education: Initiatives targeting beach users with campaigns on waste management, recycling, and environmental stewardship, thereby fostering civic responsibility.
- Climate adaptation planning: Strategic measures within municipal adaptation frameworks aimed at mitigating climate impacts on coastal infrastructures and ecosystems.

Given the scale of the climate threats, there is a pressing need to intensify adaptation strategies and strengthen the resilience of beach ecosystems and infrastructures. Such measures require significant financial investment and innovative financing instruments to ensure long-term sustainability. In this respect, societal contributions (through both public and user-based funding) are likely to play an increasingly important role in safeguarding coastal resources.

In recognition of the urgent need to repair climate-related damages, the Spanish national government launched in November 2024 a targeted financing programme for the conservation and regeneration of beaches in Málaga province, including Fuengirola. With an investment of €2.1 million, the programme has funded the repair of coastal infrastructure, sand transfers to eroded areas, and measures to conserve coastal ecosystems, all implemented in close coordination with local municipalities (Ferrary, 2024). At the local level, Fuengirola has also articulated its long-term vision for resilience through the Local Action Plan Urban Agenda 2030, which establishes objectives for climate mitigation and adaptation, risk prevention, and integrated coastal zone management. These initiatives are crucial for enhancing the municipality's adaptive capacity and ensuring its ability to cope with the multifaceted impacts of CC (Fuengirola City Council, 2024).

METHOD AND MATERIALS

The concept of Willingness to Pay (WTP) refers to the maximum monetary contribution that an individual or a collectivity is prepared to offer for the acquisition or preservation of a particular good or service. A variety of methodologies exist to estimate WTP across groups of individuals, among which the contingent valuation (CV) method has gained widespread recognition in academic research. This approach is particularly useful for assigning economic values to goods and services (such as environmental assets) that are not traded in conventional markets and therefore lack an explicit market price (Expósito et al., 2021). Although WTP estimates are inherently subjective, they provide a valuable rationale for implementing measures that may not yield direct financial returns but nonetheless enhance social welfare and environmental quality (Ramdas & Mohamed, 2014). Within the framework of CC adaptation, WTP analysis enables the valuation of preventive or adaptive measures, such as coastal conservation and adaptation measures. In addition, these estimations help assess the degree of public acceptance of specific interventions, thereby facilitating the design of more effective policies and ensuring the financial sustainability of conservation and adaptation initiatives.

Kriström and Riera (1997) describe CV as a simple method based on asking a group of people how much they would be willing to pay to obtain a particular good. In essence, this method is based on conducting surveys to potential consumers on a natural good or service, simulating a hypothetical market. In this market, the interviewer represents the supply and the respondent the demand. Through open or dichotomous questions, the aim is to determine the maximum willingness of respondents to pay to continue enjoying the benefits associated with the natural resource, thus allowing its value estimation. This approach is based on the fact that respondents are interested in reflecting their actual preferences and agree to contribute financially to the conservation of the natural resource by paying an entrance fee. Therefore, it is important to exclude protest responses (negative answers based on the discontent behavior of the respondent). In our case, the surveys were conducted in person, with the interviewer acting as a fictitious seller and the interviewee as a potential buyer. The main objective of the interviews was to identify the participants' WTP.

For this study, a double-bounded approach was used, which has been widely validated in the recent literature as a reliable method to estimate the willingness of a sample of individuals to pay for environmental services (Nesha Dushani et al., 2023). This approach asks two questions related to different monetary offers (or bids). If the respondent answers yes to the first question, they are presented with a second, higher offer. In case of a negative response, the second offer will be lower. Thus, the second question depends directly on the answer obtained in the first, which makes it possible to collect more information about the respondent compared to the single-limit approach. However, this method also requires a more complex econometric analysis due to the high amount of data obtained.

A brief description of the econometric model used in this study is presented in Expósito (2019), where the general function of the WTP can be expressed as:

$$WTP(z_i,\ u_i) = z_i'\beta + u_i$$

where z_i represents the vector of explanatory variables, including the offer amount, β is the explanatory parameter vector, and u_i corresponds to the error term.

In our specific case study, both a double-limit contingent valuation method and open-ended questions related to the maximum contribution declared by the respondents were used. Data was collected from 453 surveys conducted between December 2024 and January 2025 at various locations along the town's coastal promenade and beaches. The questionnaire was divided into four sections. A first introductory section, in which a brief explanation of the study objectives was presented to the respondents. A second section aimed to gather demographic and behavioral information from the respondents, such as gender, age, education, income, environmental concerns, beach usage, and behavioural patterns. A third section focused on questions about the respondents' awareness and perception of the CC impacts on the beach. Finally, last section is devoted to assessing the respondent's WTP to finance beach conservation and climate resilience measures based on the CV method previously presented.

As commented, our WTP estimation relies on a double-bounded dichotomous choice model, which is widely recognized in the existing literature for its reliability in environmental valuation studies (Enriquez-Acevedo et al., 2018; Expósito, 2019). A dichotomous question was posed to know if the respondents would be willing to contribute financially by paying a fee when using the beach. At this point, it is worth noting that the fee range from €0.50 to €3 was defined in a pre-test phase of 50 surveys to beach users, where 95% of surveyed people responded with amounts within this range. Firstly, the respondent was asked if he/she would be willing to pay. In case of an affirmative answer, respondents were presented with a predefined initial random bid within the defined range: 0.50, 1, 1.50, 2, 2.50, or 3 euros. If they accepted, they were offered a higher bid. On the contrary, if they reject it, they are offered a lower bid. Once the respondent was asked those two random bids, he/she was asked an open-ended question on the maximum amount the respondent would be willing to pay. This study did not use protest answers, amounting to 24% of the surveyed users. Respondents who expressed a protest to pay were asked about the reason for their refusal. Most of them answered that they already paid too much in terms of taxes and other fees and that the national/local governments should cover those beach conservation expenses. The main variables used in this study are described in Table 1.

Table 1. Variables description

Demographic variables	Description
Sex	Male (0), Female (1)
Age	<20 years (1), 20-34 years (2), 35-50 years (3), 51-65 years (4), >65 years (5)
Nationality	National (0), Foreign (1)
Income	Low income <20000 eur/year (1), Middle income between 20000 and 40000 euro/year (2), High income >40000 eur/year (3)
Educational level	Secondary education or below (1), post-compulsory secondary education (2), university graduate (3), postgraduate (4).
Expenditure (per day)	<5 Eur/day (1), between 5 and 10 Eur/day (2), >10 Eur/day (3)
Time at the beach (per day)	<1 hour (1), between 1 and 3 hours (2), >3 hours (3)
Perception of beach characteristics	
Clean water	Not important (1), important (2), very important (3)
Sand cleaning	Not important (1), important (2), very important (3)
Beach access	Not important (1), important (2), very important (3)
Comfort	Not important (1), important (2), very important (3)
Beach width	Not important (1), important (2), very important (3)
Noise	Not important (1), important (2), very important (3)
Agglomeration	Not important (1), important (2), very important (3)
Restoration	Not important (1), important (2), very important (3)
Sunbeds and umbrellas	Not important (1), important (2), very important (3)
Security	Not important (1), important (2), very important (3)
Surveillance and Rescue	Not important (1), important (2), very important (3)
Parking	Not important (1), important (2), very important (3)

continued on following page

Table 1. Continued

Demographic variables	Description
Sanitary facilities	Not important (1), important (2), very important (3)
CC perception	
Consideration of CC as a problem	Yes (1), No (0)
Knowledge of the impacts of CC on beaches	Yes (1), No (0)
Type of impact: less sand	Yes (1), No (0)
Type of impact: coastline	Yes (1), No (0)
Type of impact: higher water temperature	Yes (1), No (0)
Type of impact: damages to beach facilities	Yes (1), No (0)
Maximum WTP	Maximum amount of euros per visit

Source: Authors' own.

Unlike revealed preference methods, which infer values from observed behavior, CV methods allow researchers to capture values for goods or services not reflected in market transactions, including future-oriented or hypothetical conservation measures. This flexibility makes CV particularly suitable for assessing willingness to pay for climate adaptation or biodiversity protection, where market analogues are absent. However, CV methods also face notable limitations. Responses may be affected by hypothetical bias, strategic behavior, or difficulties in respondents' understanding of complex environmental issues. Compared to experimental or choice modeling approaches, CV provides less detailed information on trade-offs among attributes and may yield less robust welfare estimates. Moreover, results are sensitive to survey design, including question framing and payment vehicles, which can compromise comparability across studies. Despite these challenges, CV remains one of the most widely applied and policy-relevant tools for valuing non-market environmental goods. Further, the selected double-bounded dichotomous choice model has shown to be appropriate to assess respondents' WTP in recent studies (Expósito et al., 2021; Nesha Dushani et al., 2023).

RESULTS

The descriptive statistics of the variables analyzed are presented in Table 1, which reports the arithmetic mean, standard deviation, and minimum and maximum values. Beginning with the demographic characteristics, the sample reveals a slight predominance of male respondents, representing 57.17% of the total, while females account for 42.83%. In terms of age distribution, the mean value of 2.99 corresponds primarily to individuals aged between 35 and 50 years. This finding suggests that the sample is concentrated in middle adulthood, a stage often associated with higher purchasing power and greater environmental awareness compared to younger age groups. Nationality is relatively balanced, with 47.24% of participants identified as Spanish nationals and 52.76% as foreign visitors. This distribution reflects the international appeal of Fuengirola as a coastal tourist destination and offers a valuable perspective on the views of both domestic and international beach users.

Regarding socioeconomic characteristics, income levels are concentrated around the median value of 2.11, indicating that the majority of respondents fall within the middle-income group. Educational attainment in the sample is comparatively high, with an average value of 2.32. This shows that most respondents have at least completed post-compulsory education, and a substantial proportion hold a university degree or higher. The presence of a relatively well-educated sample is particularly relevant for CV studies, as higher educational attainment is often associated with greater environmental awareness and a stronger predisposition to support conservation measures financially.

Patterns of beach use also provide important contextual information. Reported daily expenditure on the beach averages 1.82, which corresponds to approximately €5 per person per day. This relatively modest level of spending highlights that beach recreation in Fuengirola is accessible to a wide variety of societal groups, but it also underscores the limited economic margins available for additional financial contributions. The average time spent on the beach per visit is 2.30 (approximately three hours). This duration suggests that visitors treat the beach as a core element of their recreational experience, but not necessarily as a full-day activity.

The analysis of perceptions regarding beach services shows clear patterns in user priorities. Sanitary facilities are ranked as the most important feature, with an average rating of 2.68. This is closely followed by lifeguard and security services (2.67) and water cleanliness (2.61). These results indicate that visitors prioritize health, safety, and hygiene (dimensions that directly influence the quality and security of the recreational experience). By contrast, services such as sunbed and umbrella rental (2.20), beach width (2.12), and parking availability (1.97) received lower ratings. This suggests that while comfort and accessibility are relevant, they are not decisive in shaping overall user satisfaction.

Perceptions of CC are strong among respondents. Nearly unanimous agreement exists regarding the impacts of CC on the beach, with an average score of 0.98, showing that respondents recognize the relevance of this global phenomenon for coastal areas. Awareness of specific CC impacts is also relatively high (mean value 0.85), although the variability in responses points to uneven knowledge levels among users. The most frequently cited impact is the increase in water temperature, identified by 40.4% of respondents. This is followed by reduced sand availability (30%), shoreline retreat (22%), and damage to beach facilities (17%). Notably, more than one-third of respondents (35.98%) reported not perceiving any visible changes, which suggests the need for targeted awareness campaigns to increase recognition of the ongoing and future risks of CC for coastal environments.

Finally, the analysis of respondents' WTP for beach conservation and climate adaptation measures reveals an average maximum contribution of €1.37 per visit. Although this amount may seem modest, it acquires significant weight when multiplied by the volume of visitors to a tourist destination such as Fuengirola. Official statistics show that the town of Fuengirola received more than 500,000 visitors in 2024 (Spanish National Institute of Statistics). Therefore, preliminary results underscore the potential for mobilizing financial resources through user contributions, provided that adequate mechanisms are put in place to align visitor willingness with sustainable coastal management objectives.

Table 2. Descriptive statistics

Variables	Mean value	Standard deviation	Minimum value	Maximum value
Demographic variables				
Sex	0.4282	0.4953	0	1
Age	2.9933	1.0076	1	5
Nationality	0.4724	0.4997	0	1
Income	2.1125	0.5388	1	3
Educational level	2.3245	0.8191	1	4
Spending (per day)	1.8211	0.6458	1	3
Time at the beach (per day)	2.3024	0.6476	1	3
Perception of beach characteristics				
Clean water	2.6114	0.4969	1	3
Sand cleaning	2.5209	0.5045	1	3
Beach access	2.4194	0.5880	1	3
Comfort	2.5452	0.5116	1	3
Beach width	2.1236	0.7087	1	3
Noise	2.3730	0.6059	1	3
Agglomeration	2.4083	0.5750	1	3
Restoration	2.3002	0.6189	1	3
Sunbeds and umbrellas	2.2008	0.6563	1	3

continued on following page

Table 2. Continued

Variables	Mean value	Standard deviation	Minimum value	Maximum value
Security	2.6688	0.4804	1	3
Surveillance and Rescue	2.6710	0.4796	1	3
Parking	1.9713	0.7326	1	3
Sanitary facilities	2.6754	0.4734	1	3
Perceptions of CC impacts				
Consideration of CC as a problem	0.9823	0.1318	0	1
Knowledge of the impacts of CC on beaches	0.8476	0.3597	0	1
Type of impact: less sand	0.3024	0.4598	0	1
Type of impact: coastline	0.2163	0.4122	0	1
Type of impact: increased water temperature	0.4039	0.4912	0	1
Type of impact: damages to beach facilities	0.1655	0.3720	0	1
Maximum WTP	1.3704	1.4878	0	10

Source: Authors' own.

Table 3 presents the correlations between selected socioeconomic and users' perception variables, detailing the correlation coefficients and their corresponding statistical significance values (p-values). Income exhibits a statistically significant positive correlation with nationality ($r = 0.25$; $p < 0.00$), indicating that income levels vary depending on the respondents' origin, with foreign participants reporting higher income levels than Spanish nationals. However, no statistically significant relationships are identified between income and other variables such as gender or age. Educational attainment is positively and significantly correlated with nationality ($r = 0.32$; $p < 0.00$), suggesting that foreign respondents tend to have higher levels of education than their Spanish counterparts. Furthermore, a positive correlation is observed between educational level and income ($r = 0.29$; $p < 0.00$), as well as with age ($r = 0.10$; $p < 0.05$).

These findings indicate that older individuals and those with higher incomes are more likely to have attained higher levels of education. Daily expenditure exhibits a strong positive correlation with nationality ($r = 0.43$; $p < 0.00$), confirming that foreign tourists tend to spend more than national visitors. Additionally, significant positive relationships are identified between expenditure patterns and both income ($r = 0.26$; $p < 0.00$) and educational level ($r = 0.28$; $p < 0.00$). These results reinforce the fact that individuals with higher socioeconomic status tend to spend more money on beach-related services and activities. The amount of daily time spent at the beach shows a significant negative correlation with gender (r = –0.12; p = 0.01), suggesting that men generally spend more time on the beach than women. Moreover, significant positive correlations are identified with income (r = 0.16; p < 0.00) and educational level (r = 0.18; p < 0.00), indicating that individuals with higher education and income levels tend to spend more time at the beach. The maximum WTP shows significant positive correlations with nationality (r = 0.21; p < 0.00), income (r = 0.17; p < 0.00), and educational level (r = 0.17; p < 0.00). These findings indicate that foreign visitors and individuals with higher income and education levels are more willing to contribute financially to the conservation of coastal areas. Furthermore, daily expenditure on the beach exhibits a significant positive correlation with the maximum WTP (r = 0.31; p < 0.00), suggesting that those visitors who spend more money during their beach visits are also more inclined to support environmental conservation efforts.

Table 3. Correlation coefficients

Variables	Sex	Age	Nationality	Income	Education	Expenditure	Visit length
Income	0.01	0.07	0.25	1.00			
p-value	0.84	0.16	0.00				
Education	-0.07	0.10	0.32	0.29	1.00		
p-value	0.13	0.03	0.00	0.00			
Expenditure	-0.09	0.07	0.43	0.26	0.28	1.00	
p-value	0.07	0.12	0.00	0.00	0.00		
Visit length	-0.12	0.04	0.15	0.16	0.18	0.20	1.00
p-value	0.01	0.39	0.00	0.00	0.00	0.00	
Max WTP	0.05	0.00	0.21	0.17	0.17	0.31	0.02
p-value	0.32	0.92	0.00	0.00	0.00	0.00	0.71

Source: Authors' own.

Our analysis reveals that socioeconomic and attitudinal variables significantly influence respondents' WTP and stated preferences. Table 4 synthesizes the average maximum WTP by nationality, gender, age, income, educational level, environmental concern, daily time spent at the beach, daily expenditure, and perceived CC impacts, thereby allowing for a more nuanced interpretation of the data.

Nationality emerges as one of the strongest differentiating factors. Spanish respondents report an average WTP of €1.07, while foreign visitors are willing to pay €1.70 per visit. This €0.63 differential reflects both the temporary nature of foreign tourists' stay (likely perceived as a valuable and time-limited experience) and their comparatively higher economic capacity. Moreover, international tourists may attach greater symbolic value to Mediterranean beaches as cultural and natural attractions, which enhances their readiness to contribute to conservation. This result aligns with prior studies suggesting that non-residents are often more willing to contribute to the preservation of environmental amenities they perceive as unique or scarce (Enriquez-Acevedo et al., 2018).

Gender differences are also evident. Women's average WTP (€1.45) exceeds that of men (€1.31). While the difference is modest, it is consistent with literature indicating that women frequently display stronger pro-environmental attitudes and greater concern for communal well-being, which translates into higher valuation of conservation initiatives (Zelezny et al., 2000). This pattern suggests that communication strategies aimed at fostering contributions might resonate particularly well with female visitors, though men's slightly lower WTP still demonstrates broad support across genders.

The age distribution of WTP presents a non-linear pattern. The youngest group, those under 20 years old, report the lowest contribution (€1.21), reflecting their limited financial capacity. WTP increases among respondents aged 20-34 (€1.39) and 35-50 (€1.35), corresponding to life stages associated with higher income and consumption capacity. The highest value is found among individuals aged 51-65 (€1.46), who may combine financial stability with a heightened preference for quality recreational services. Respondents above 65 years show a decline (€1.23), though still higher than the youngest cohort, possibly reflecting fixed incomes and more conservative spending behavior. This pattern suggests that middle-aged and pre-retirement groups represent the segment most inclined to support conservation financially, while youth and elderly populations might require alternative engagement strategies.

Income also exerts a strong influence on the respondents' WTP. High-income visitor report an average contribution of €2.00, substantially above low- and middle-income groups (€1.33 and €1.19, respectively). These results are consistent with the general principle of income elasticity in environmental valuation: as disposable income rises, so does the capacity and willingness to contribute to public goods (Carson & Hanemann, 2005). Interestingly, middle-income respondents express lower WTP than the low-income group, possibly reflecting higher expenditure constraints and a greater need to prioritize essential spending.

Educational attainment further differentiates WTP levels. Respondents with secondary education or below report the lowest value (€1.03), while post-compulsory secondary education slightly increases the estimate (€1.30). University degree holders contribute marginally more (€1.39). The most striking difference is observed among postgraduate respondents, whose average WTP (€2.49) is more than double that of respondents with lower education. These differences underscore the role of education in shaping environmental awareness, pro-conservation attitudes, and the perceived legitimacy of contributing financially to conservation initiatives.

The role of environmental concern is equally noteworthy. Participants who describe themselves as "not concerned" about environmental issues contribute the lowest average amount (€1.06). Those who are "worried" report a modest increase (€1.19), while "very worried" respondents show the highest WTP (€1.64). These significant differences highlight the close relationship between risk perception and economic support for beach conservation measures, suggesting that awareness campaigns can directly increase financial contributions by improving the understanding of climate threats.

Visitor behavior, measured by daily time spent at the beach and daily expenditure, also provides interesting insights. The duration of stay shows only a limited effect: respondents staying less than one hour and those exceeding three hours display similar WTP values (€1.42), slightly above the group staying between one and three hours (€1.31). This indicates that frequency of use alone does not strongly predict WTP, and the relationship may be shaped by heterogeneous user profiles. By contrast, daily expenditure exhibits a clear and robust positive correlation. Visitors spending less than €5 per day express the lowest WTP (€0.82), while those spending between €5 and €10 increase their contribution (€1.46). The highest WTP is reported by respondents spending more than €10 daily (€2.29), confirming that willingness to contribute scales with overall recreational spending and perceived quality of experience.

Finally, perception of CC impacts proves to be a critical determinant. Respondents who had noticed changes to the beach environment report an average WTP of €1.43, compared to €1.02 among those who had not. This demonstrates that personal observation of environmental degradation strengthens the perceived urgency of conservation and enhances visitors' readiness to contribute financially. The difference is substantial enough to suggest that communication strategies highlighting visible impacts, such as coastal erosion, sand loss, or water temperature increases, could significantly boost support for beach conservation measures.

Taken together, these results underscore the importance of socio-demographic, attitudinal, and behavioral factors in shaping visitors' WTP. Nationality, education, income, and environmental concern stand out as the strongest predictors, while daily expenditure provides an additional behavioral indicator closely tied to economic commitment. By contrast, time spent at the beach and age show weaker or more complex associations. These findings provide valuable guidance for designing financing mechanisms that are socially equitable, economically viable, and sensitive to the diverse profiles of beach users.

Table 4. Average WTP by respondents' groups

Groups	Average maximum WTP (Euro)
Nationality	
National	1.07
Foreign	1.70
Sex	
Male	1.31
Female	1.45
Age	
< 20	1.21
20 - 34	1.39
35 - 50	1.35
51 - 65	1.46
> 65	1.23
Income	
Low	1.33
Middle	1.19
High	2.00
Educational level	
Secondary education or below	1.03
Post-compulsory secondary education	1.30
University degree	1.39
Postgraduate	2.49
Environmental concern	
Nothing	1.06
Worried	1.19
Very worried	1.64
Daily time at the beach	
Less than 1 hour	1.42
Between 1 and 3 hours	1.31
More than 3 hours	1.42
Daily expenditure on the beach	
Less than 5 euros	0.82
Between 5 and 10 euros	1.46
More than 10 euros	2.29
Changes noticed on the beach due to CC	

continued on following page

Table 4. Continued

Groups	Average maximum WTP (Euro)
No	1.02
Yes	1.43

Source: Authors' own.

Finally, our double-bounded dichotomous choice model provides an estimated average WTP of €1.06, with an associated standard error of 0.55. This coefficient is statistically significant at a significance level of 10% (p=0.06). The coefficient value (1.06) indicates that, on average, respondents would be willing to pay an amount close to this figure to finance measures related to the conservation of beaches and their related ecosystems in a context of increasing negative impacts of CC. This modest value represents a significant source of financial resources for the municipality, as the number of visits to Fuengirola beaches is significant. Although there are no official statistics on the number of visits (or users) to the local beaches, the number of tourists visiting the municipality of Fuengirola amounted to 526,629 in 2024 (Spanish National Institute of Statistics).

POLICY IMPLICATIONS AND RECOMMENDATIONS

Coastal municipalities across the Mediterranean face increasing CC pressures, such as sea-level rise, coastal erosion, biodiversity loss, and increasingly frequent extreme weather events. Although the study focuses on the town of Fuengirola, its implications extend well beyond this municipality. Its reliance on tourism revenues, the heterogeneous mix of domestic and international visitors, and its acute exposure to climate-related risks are shared by many Mediterranean and European coastal towns. As such, the findings can serve as an important reference point for municipal, regional, and supranational efforts to design effective and socially responsive policies for beach management and climate resilience.

Our analysis provides a nuanced picture of visitor profiles, perceptions, and attitudes towards financing conservation. Based on the findings, multiple policy implications and recommendations can be extracted. Particularly, awareness on CC impacts emerges as a significant factor that determine a positive WTP by beach users. Nearly all respondents recognize CC as a critical problem for the conservation coastal areas, yet the degree of observed impacts varies. Increases in water temperature are the most commonly cited effect, followed by reductions in sand availability and shoreline retreat. Notably, those who claim to have noticed environmental changes on the beaches express a higher willingness to contribute, while those who report no visible changes declare lower WTP values. This pattern reveals the extent to which direct perceptions of environmental degradation translate into greater readiness to support adaptation and conservation financially. Additionally, the CV model offers an average WTP of €1.06 per visit. While this figure may appear modest on an individual scale, its aggregate significance becomes evident when multiplied across the millions of visitors who frequent Mediterranean coastal destinations annually. Such evidence suggests that, although visitor contributions cannot replace public investment, they can represent a complementary and reliable stream of revenue for municipalities seeking to fund conservation and adaptation measures.

Policy Implications

The municipality of Fuengirola, located on Spain's Costa del Sol, offers a particularly illustrative case in which environmental vulnerability intersects with the socio-economic significance of beach tourism. The results of this study carry important implications for coastal policy design. First, the positive WTP values confirm that user-based financing mechanisms are viable, particularly in tourist-driven economies where even small contributions, if broadly collected, can generate meaningful revenues. Options might include voluntary donation schemes, visitor fees incorporated into tourism services, or earmarked levies on specific activities such as parking or sunbed rentals. To succeed, however, such measures must be accompanied by transparency and clear communication, ensuring that visitors can see a direct link between their contributions and visible improvements in coastal management (Booth et al., 2022).

The heterogeneity observed across demographic groups further suggests that financing schemes should be carefully designed to reflect different user profiles. Foreign tourists, who both perceive higher value and have greater disposable income, are more willing to contribute than local residents. Similarly, individuals with higher education and income levels express greater readiness to support conservation measures. This indicates the potential for differentiated mechanisms, such as modest surcharges incorporated into tourist accommodation or premium services that channel part of their revenue directly into conservation funds. Such approaches would prevent undue burden on residents and lower-income groups, while still harnessing the financial potential of higher-spending visitors.

The close link between environmental awareness and WTP points to the importance of education and communication strategies. Visitors who recognize CC impacts are significantly more willing to contribute, suggesting that municipalities should invest in campaigns that make the risks and consequences of coastal degradation visible. Information panels at beach entrances, digital campaigns, or participatory activities such as guided environmental walks could all reinforce the connection between individual contributions and the broader sustainability of coastal ecosystems.

Finally, the integration of visitor contributions into broader adaptation planning frameworks is essential. Municipal plans, such as Fuengirola's "Local Action Plan Urban Agenda 2030", already identify CC adaptation as a strategic objective, but they often lack robust financing instruments. Incorporating user-based funding as a complement to national and EU allocations would not only diversify financial resources but also enhance the legitimacy of adaptation policies by demonstrating that they are co-financed by the users who benefit from healthy beaches.

In designing these mechanisms, equity considerations must remain central. Residents frequently express resistance to new charges when they perceive that their taxes already cover beach management. To mitigate this, municipalities should clearly communicate that visitor contributions are supplementary, not replacements, for public financing. Moreover, exemptions or reduced rates could be considered for local households or vulnerable groups, ensuring that adaptation measures do not inadvertently generate social inequalities.

Recommendations for Coastal Municipalities

The evidence from this study highlights the opportunities and challenges for municipalities in designing effective policies to mobilize visitor contributions for climate adaptation and beach conservation. Further, these policies should balance different aspects regarding economic viability, social equity, and environmental effectiveness of alternative measures.

As a previous step, municipalities should prioritize improving communication, transparency, and stakeholder trust. The findings show that willingness to pay is higher among visitors who are aware of the CC impacts. This underscores the importance of education and awareness campaigns. Local governments could introduce information panels at beach access points explaining how CC is affecting local coastlines and what specific conservation measures are underway. Transparent accounting mechanisms should accompany any financing initiative, such that visitors can clearly see how their contributions are used (e.g., through annual reports or digital dashboards showing investments in sand nourishment, dune restoration, or lifeguard infrastructure). Establishing this credibility is crucial for public acceptance of new financing tools (Mycoo & Gobin, 2013; Griggs & Reguero, 2021).

Regarding alternative schemes for collecting visitor contributions, several models could be considered. A voluntary "coastal stewardship contribution" could be added to accommodation bills, similar to tourist eco-taxes already applied in parts of Spain and Italy. Alternatively, municipalities might earmark a small portion of revenues from existing beach services, such as parking fees, umbrella rentals, or concession licenses, to conservation funds. Additionally, these mechanisms should be designed to reflect the socio-economic heterogeneity revealed in the results. Higher-spending foreign tourists, for example, might contribute more through accommodation surcharges, while local residents could benefit from exemptions or reductions.

With the aim to enhance public acceptance and environmental effectiveness, municipalities should embed these financing instruments within comprehensive climate adaptation strategies (Cunha-e-Sá et al., 2025). Visitor contributions, while important, cannot replace public investment; rather, they should complement public funds, national climate adaptation programs, and private sector participation. Over time, a dedicated coastal adaptation fund could be created, pooling resources from multiple sources, including visitor contributions, public funds, and partnerships with local businesses. This would allow for larger-scale interventions, such as the restoration of dune systems, construction of nature-based coastal defenses, and investment in monitoring technologies. In the long term, visitor contributions should be embedded within broader strategies for sustainable coastal governance. Nature-based solutions, such as dune restoration, wetland conservation, and seagrass bed protection, offer cost-effective and ecologically sound alternatives to hard infrastructure. Investments in predictive modeling and monitoring technologies can improve municipalities' ability to anticipate climate impacts and plan accordingly.

Another central lesson from the findings is that users' WTP is not uniform across societal groups. Designing equitable and effective financing schemes therefore requires governance approaches that engage stakeholders at multiple levels (Onofrio et al., 2025). At the local level, municipal governments should involve residents, business owners, and civil society organizations in the design of contribution mechanisms. Participatory forums or stakeholder workshops could help identify acceptable forms of contribution and build legitimacy around policy measures. Hotels, restaurants, and beach concessionaires could play a critical role as intermediaries in collecting small voluntary contributions from tourists. At the regional and national level, coordination is necessary to avoid fragmentation. If each municipality designs its own fee structure independently, visitors may perceive inconsistency or unfairness, potentially undermining compliance. Regional frameworks, led by provincial authorities or autonomous communities, could provide guidance and harmonization. At the national scale, ministries of environment and tourism should integrate visitor contributions into broader climate adaptation strategies, ensuring complementarity with existing public financing instruments. At the European level, the alignment of local initiatives with EU strategies is essential. The European Green Deal, the EU Adaptation Strategy, and the EU Blue Economy strategy already prioritize coastal resilience. Demonstrating that municipalities are leveraging visitor contributions can help secure co-financing from EU funds, creating a virtuous circle of local initiative and supranational support.

Beyond financing, the study highlights the critical role of environmental awareness. Respondents who had directly perceived CC impacts reported significantly higher WTP. This finding suggests that municipalities must invest not only in infrastructure but also in communication and education. Public awareness campaigns could emphasize the direct connection between contributions and tangible outcomes: cleaner beaches, healthier ecosystems, and safer recreational spaces. Schools and universities can be involved in citizen science projects, such as monitoring beach erosion or water quality, linking younger generations to conservation goals. Digital communication strategies, including mobile apps or interactive maps, could engage visitors in real time, showing how contributions are being used to improve the coastal environment.

Other measures, such as the construction or reinforcement of defensive structures (e.g., dikes, retaining walls and other physical infrastructures), provide physical barriers against waves. Despite having a high financial cost, they help to prevent damage and protect coastal areas from flooding and erosion. Furthermore, it is advisable to invest in research and innovation initiatives aimed to better understand coastal processes. The development of new technologies for climate adaptation and mitigation can lead to more effective and sustainable solutions to the challenges posed by CC on our coastal ecosystems. Along these lines, the improvement of predictive models that can accurately estimate the evolution of coastlines and the CC impacts would be very useful, thus allowing for more effective planning and implementation of appropriate adaptation measures (Vitousek et al., 2017).

CONCLUSION

Coastal municipalities must address the dual challenge of safeguarding ecosystems and sustaining tourism-based economies under accelerating CC and its related extreme weather events. Evidence from CV studies offers a powerful tool for designing funding mechanisms that combine public investment, visitor contributions, and private sector participation. The path forward requires not only financial innovation but also participatory governance, multi-level coordination, and a long-term vision balancing sustainability and equity principles. The integration of visitor contributions into climate adaptation strategies can help secure the resilience of coastal zones, ensuring that beaches remain valuable economic and ecological assets for future generations.

The results of this study provide strong evidence that visitors to Mediterranean beaches are willing to contribute financially to conservation measures, although this might be shaped by socio-economic and attitudinal factors. For coastal municipalities, like Fuengirola, the policy challenge lies in translating this willingness into concrete financing mechanisms that are transparent, equitable, and socially accepted.

Results have shown that socioeconomic, demographic, and attitudinal factors influence visitors' WTP, highlighting the relevance of understanding user preferences in formulating public policies and coastal management strategies. Higher income, environmental concerns, and frequent beach visitors exhibited greater WTP estimates for conservation measures. Foreign tourists have been identified by their higher WTP, compared to national visitors. Similarly, women have demonstrated slightly higher WTP estimates than men. Higher-income and higher-educated respondents exhibited greater WTP estimates. Finally, those respondents who expressed strong concern for CC and its impacts on coastal areas were more inclined to contribute financially with a significantly higher WTP. These findings highlight the importance of tailoring coastal management policies to different demographic groups, ensuring equitable contributions to conservation efforts.

From a policy perspective, this study underlines the need to involve society in decision-making related to coastal management, ensuring that the measures implemented are perceived as beneficial and aligned with their social values and priorities. Likewise, the integration of environmental education and awareness programs could be a key tool to increase awareness of the importance of CC adaptation and conservation measures to protect beaches and coastal ecosystems, as well as to mitigate possible resistance to financing by specific population sectors. In addition, the positive visitors' WTP for the implementation of beach conservation measures in the face of the negative impacts of CC opens up new forms of financing for local entities, which would result in more financial resources to improve the climate resilience of beaches, their facilities and their associated ecosystems. In a context marked by the increasing frequency and intensity of extreme weather events, the design of sustainable and economically viable policies for protecting beaches is crucial. Ultimately, the results of this study constitute a solid basis for future research and policy development that combines environmental sustainability, social participation, and economic equity.

The conservation measures implemented by the Fuengirola municipality represent a significant effort to protect and improve beaches and coastal ecosystems in the face of the challenges posed by the impacts of CC. However, these must, in turn, be accompanied by more extensive and forward-looking actions that allow us to anticipate CC scenarios and, at the same time, ensure more sustainable and efficient coastal management. Additionally, coastal conservation actions should be designed and implemented in collaboration with societal stakeholders (e.g., civil representatives, environmental organisations, tourism sector) with the aim to ensure sustainable and equitable solutions. Further, visitor monetary contributions should not be considered an unique instrument to finance conservation measures, but as a complement to public funds, national climate adaptation programs, and private sector participation.

Finally, it is worth noting that the case study analised in this chapter provides evidence of the necessity of innovative financing mechanisms for coastal conservation. The estimated WTP of beach users offers valuable insights for policymakers to design climate adaptation strategies and financing mechanisms that facilitate the involvement of society in the protection of coastal resources. Further, the implementation of public participation schemes and environmental education programs can enhance awareness and financial support for coastal protection measures. Given the increasing frequency of extreme weather events and their negative impacts on coastal areas, future research should explore how alternative funding instruments might include visitors contributions in a sustainable and equitable way to ensure the long-term sustainability of the coastal tourism industry.

REFERENCES

Alves, B., Rigall-I-Torrent, R., Ballester, R., Benavente, J., & Ferreira, Ó. (2015). Coastal erosion perception and willingness to pay for beach management (Cadiz, Spain). *Journal of Coastal Conservation, 19*(3), 269–280. DOI: 10.1007/s11852-015-0388-6

Birdir, S., Ünal, Ö., Birdir, K., & Williams, A. T. (2013). Willingness to pay as an economic instrument for coastal tourism management: Cases from Mersin, Turkey. *Tourism Management, 36*, 279–283. DOI: 10.1016/j.tourman.2012.10.020

Booth, H., Mourato, S., & Milner-Gulland, E. J. (2022). Investigating acceptance of marine tourism levies, to cover the opportunity costs of conservation for coastal communities. *Ecological Economics, 201*, 107578. DOI: 10.1016/j.ecolecon.2022.107578

Choi, E. C., Lee, J. S., & Chang, J. I. (2021). Willingness to pay for the prevention of beach erosion in Korea: The case of Haeundae beach. *Marine Policy, 132*, 104667. DOI: 10.1016/j.marpol.2021.104667

Cunha-e-Sá, M. A., Dietrich, T., Faria, A., Nunes, L. C., Ortigão, M., Rosa, R., & Vieira-da-Silva, C. (2025). Willingness to pay for nature protection: Crowdfunding as a payment mechanism. *Environmental and Resource Economics, 88*(2), 367–396. DOI: 10.1007/s10640-024-00933-3

Do, Ó. A., & Seiz R. (2021). Effects of climate change in the Iberian Peninsula. Less rain and more uncertainty for the rivers. ANP/WWF and WWF Spain https://wwfes.awsassets.panda.org/downloads/efectos_cc_en_la_peninsula__2_.pdf?57520/Informe-sobre-los-efectos-del-cambio-climatico-en-la-Peninsula-Iberica

Enriquez-Acevedo, T., Botero, C. M., Cantero-Rodelo, R., Pertuz, A., & Suarez, A. (2018). Willingness to pay for Beach Ecosystem Services: The case study of three Colombian beaches. *Ocean and Coastal Management, 161*, 96–104. DOI: 10.1016/j.ocecoaman.2018.04.025

Expósito, A. (2019). Valuing Households' Willingness to Pay for Water Transfers from the Irrigation Sector: A Case Study of the City of Seville (Southern Spain). *Sustainability (Basel), 11*(24), 6982. DOI: 10.3390/su11246982

Expósito, A., Espinosa, M., & Villa-Damas, A. (2021). Valuing visitor willingness to pay for urban green space conservation: Case of Maria Luisa Park in Seville, Spain. *Journal of Urban Planning and Development, 147*(3), 05021020. DOI: 10.1061/(ASCE)UP.1943-5444.0000700

Ferrary, M. (2024, November 25). The government begins the maintenance of the beaches along the entire coast of Malaga. La Opinión de Málaga. https://www.laopiniondemalaga.es/malaga/2024/11/25/gobierno-trabajos-mantenimiento-playas-malaga-112025799.html

Freeman, A. M., Herriges, J. A., & Kling, C. L. (2014). *The Measurement of Environmental and Resource Values: Theory and Methods* (3rd ed.). Routledge. DOI: 10.4324/9781315780917

Fuengirola City Council. (2023, March 22). Fuengirola achieves the 'S for Sustainability' certificate, from the Institute of Tourism Quality of Spain, on all its beaches. Fuengirola. https://www.fuengirola.es/fuengirola-logra-el-certificado-s-de-sostenibilidad-del-instituto-de-calidad-turistica-de-espana-en-la-totalidad-de-sus-playas/

Fuengirola City Council. (2024). Sustainability Plan. Department of Beaches. https://www.fuengirola.es/wp-content/uploads/2024/07/PLAN-SOSTENIBILIDAD-OGP.pdf

Griggs, G., & Reguero, B. G. (2021). Coastal adaptation to climate change and sea-level rise. *Water (Basel)*, *13*(16), 2151. DOI: 10.3390/w13162151

Hanley, N., & Barbier, E. (2009). *Pricing Nature: Cost-Benefit Analysis and Environmental Policy*. Edward Elgar.

IPCC. (2019). Resumen para responsables de políticas. https://www.ipcc.ch/site/assets/uploads/sites/2/2019/09/IPCC-Special-Report-1.5-SPM_es.pdf

Kontogianni, A., Damigos, D., Tourkolias, Ch., Vousdoukas, M., Velegrakis, A., Zanou, B., & Skourtos, M. (2014). Eliciting beach users' willingness to pay for protecting European beaches from beachrock processes. *Ocean and Coastal Management*, *98*, 167–175. DOI: 10.1016/j.ocecoaman.2014.06.019

Kriström, B., & Riera, P. (1997). Contingent valuation method. Applications to the Spanish rural environment. *Economía Agraria*, *179*, 133–166.

Laghrissi, C. (2024, April 18). A NASA map warns of the "worrying" rise in sea level in Malaga. La Opinión de Málaga. https://www.laopiniondemalaga.es/malaga/2024/04/18/mapa-nasa-alerta-preocupante-crecida-nivel-mar-malaga-101234915.html

Liu, J., Zhao, Y., & Jang, S. C. (2021). Environmental perceptions and willingness to pay for preservation: Evidence from beach destinations in China. *International Journal of Tourism Research*, *23*(5), 792–804. DOI: 10.1002/jtr.2442

Mitchell, R. C., & Carson, R. T. (1989). *Using Surveys to Value Public Goods: The Contingent Valuation Method*. Resources for the Future Press.

Mycoo, M. A., & Gobin, J. F. (2013). Coastal management, climate change adaptation and sustainability in small coastal communities: Leatherback turtles and beach loss. *Sustainability Science*, *8*(3), 441–453. DOI: 10.1007/s11625-013-0212-x

Nesha Dushani, S., Aanesen, M., & Armstrong, C. W. (2023). Willingness to pay for mangrove restoration to reduce the climate change impacts on ecotourism in Rekawa coastal wetland, Sri Lanka. *Journal of Environmental Economics and Policy*, *12*(1), 19–32. DOI: 10.1080/21606544.2022.2065364

Onofrio, F., Rodella, I., & Gilli, M. (2025). Navigating WTP disparities: A study of tourist and resident perspectives on coastal management. *Frontiers in Environmental Economics*, *4*, 1497532. DOI: 10.3389/frevc.2025.1497532

Pérez-Juan, C. M., Prieto Fustes, L., Salvo Tierra, A. E., Navarro Jurado, E., Teixidor Costa, A., Danelutti, C., Iglesias Saugar, J., Wahlberg Macías, N., Jiménez Arce, J. A., & Velasco Martínez, L. C. (2022). *Tourism and the climate crisis on the Costa del Sol: Preliminary diagnosis and proposals for action. Gland, Switzerland: IUCN and Málaga*. University of Málaga.

Ramdas, M., & Mohamed, B. (2014). Impacts of tourism on environmental attributes, environmental literacy and willingness to pay: A conceptual and theoretical review. *Procedia: Social and Behavioral Sciences*, *144*, 378–391. DOI: 10.1016/j.sbspro.2014.07.307

Rejón, R. (2024, June 6). Climate change eats up 45 metres of Spanish coastline weakened by construction in five years. ElDiario.es. https://www.eldiario.es/sociedad/cambio-climatico-come-cinco-anos-45-metros-costa-espanola-debilitada-construccion_1_11420555.html

Rodella, I., Madau, F., Mazzanti, M., Corbau, C., Carboni, D., Utizi, K., & Simeoni, U. (2019). Willingness to pay for management and preservation of natural, semi-urban and urban beaches in Italy. *Ocean and Coastal Management*, *172*, 93–104. DOI: 10.1016/j.ocecoaman.2019.01.022

Schuhmann, P. W., Bangwayo-Skeete, P., Skeete, R., Seaman, A. N., & Barnes, D. C. (2024). Visitors' willingness to pay for ecosystem conservation in Grenada. *Journal of Sustainable Tourism*, *32*(8), 1644–1668. DOI: 10.1080/09669582.2023.2259119

Vitousek, S., Barnard, P. L., & Limber, P. (2017). Can beaches survive climate change? *Journal of Geophysical Research. Earth Surface*, *122*(4), 1060–1067. DOI: 10.1002/2017JF004308

KEY TERMS AND DEFINITIONS

Climate Change (CC): Long-term alterations in global or regional climate patterns, largely driven by human activities such as greenhouse gas emissions, resulting in shifts in temperature, precipitation, and the frequency of extreme weather events.

Beach Conservation: The management and protection of coastal areas and ecosystems to maintain their ecological integrity, recreational value, and resilience against natural and human-induced pressures.

Willingness to Pay (WTP): The maximum monetary amount an individual is prepared to contribute for a good, service, or policy, often used to measure the perceived value of non-market environmental resources.

Contingent Valuation (CV): A survey-based economic method that estimates the value individuals assign to non-market goods or services by eliciting their stated willingness to pay under hypothetical scenarios.

Financing Instruments: Mechanisms, tools, or strategies (such as taxes, fees, subsidies, or public–private partnerships) used to mobilize financial resources for the implementation of policies, projects, or conservation measures.

Chapter 8
Türkiye's Water Future: Demand Forecasting and Sustainable Strategies

Beyza Güdek
https://orcid.org/0000-0002-7432-9234
Karadeniz Technical University, Turkey

Ali Gökhan Gölçek
https://orcid.org/0000-0002-7948-7688
Niğde Ömer Halisdemir University, Turkey

ABSTRACT

This study analyzes future water stress trends in Turkey and identifies the key socioeconomic and environmental drivers shaping this trajectory. Using data from 1992–2022, a multivariate polynomial regression model was developed with sectoral water use indicators, population, economic growth, CO_2 emissions, and drought measures. The model effectively captures nonlinear dynamics and explains most of the variation in water stress. Results show that population growth, agricultural withdrawals, rising GDP, and CO_2 emissions intensify water stress, while water use efficiency indicators exert a mitigating influence. Projections suggest that Turkey's water stress will continue to rise through 2050, potentially reaching critical levels under current trends. The study underscores three policy priorities: improving agricultural efficiency, reducing urban water losses and expanding reuse, and transforming water-intensive industrial processes. A holistic water management framework is essential for strengthening Turkey's resilience to escalating water stress.

DOI: 10.4018/979-8-3373-1777-9.ch008

INTRODUCTION

Increasing global water demand has become one of the most significant environmental problems threatening water security by placing increasing pressure on limited freshwater resources. Population growth, urbanization, economic growth, and changing consumption patterns are straining the regenerative capacity of freshwater resources; in addition, climate change, through irregular precipitation patterns, rising temperatures, and increased evapotranspiration, is making the water cycle more vulnerable. These multidimensional pressures demonstrate that water stress is not merely a physical water scarcity but also a complex phenomenon shaped by socioeconomic processes, technological capacity, and environmental governance. The literature indicates that water demand has more than doubled, particularly in the post-1960 period, driven by irrigation, and that many regions are approaching critical thresholds (Wada et al., 2011). In regions vulnerable to climate change, such as the Mediterranean basin and the Middle East, decreased precipitation, increased temperatures, and increased evaporation rates are creating critical areas where water stress is spatially concentrated (Chenoweth et al., 2011).

Türkiye is one of the countries where these global trends are clearly observed due to climate vulnerability, demographic pressure, and increasing sectoral water demand. Long-term hydrological records reveal significant irregularities in both seasonal and spatial rainfall distribution in Türkiye, and a notable increase in the frequency of drought events since 1980 (Yetik et al., 2023). The expansion of agricultural irrigation, increased industrial production, and rapid urbanization in the country are accelerating the rise in water demand, pushing the water budget to an unsustainable structure, especially in closed and semi-closed basins (Çolak et al., 2022). Studies showing that increasing evapotranspiration in the Mediterranean region exacerbates the vulnerability of water supply also confirm that Türkiye is shifting towards a more climatically stressed regime (Unnisa et al., 2023). In this context, the historical trend of increasing water stress in Türkiye emerges as a structural problem encompassing both climatic and socio-economic pressures.

As part of the Sustainable Development Goals, countries need to increase water use efficiency, reduce sectoral water stress, and strengthen ecosystem-based water management approaches. However, the literature shows that water stress is affected not only by current demand levels but also by multidimensional determinants such as sectoral water use, economic growth, CO2 emissions, population growth, industrial activity, and drought indicators (Vanham et al., 2018). Therefore, forecasting water stress for the future is crucial for policymakers to develop sectoral planning, water allocation, infrastructure investments, and climate adaptation strategies. In this context, the need for advanced methods that allow for the representation of nonlinear relationships in water stress prediction models is gaining importance.

Polynomial regression models are widely used in the literature due to their success in capturing long-term trends in environmental time series (De Luna, 1998; West & Harrison, 1989). Machine learning-based approaches have been found to be particularly effective in capturing sudden changes in water demand and multidimensional relationship networks (Brentan, Meirelles, et al., 2017). These methodological advances allow water stress predictions to become more precise and more directly integrated into policymaking.

This study develops a multivariate polynomial regression model that generates water stress projections for 2023–2050 for Türkiye using socioeconomic, environmental, and sectoral indicators for the period 1992–2022. The model holistically addresses the structural determinants of water stress with a broad set of variables consisting of various indices. Thus, the study analyzes the multidimensional nature of water stress, shaped not only by physical supply but also by economic structure, climatic pressures, sectoral demand, and productivity indicators. In this context, the primary objective of the study is to predict with high accuracy the level of water stress that Türkiye will face in the future and to analytically reveal the determinants shaping these trends. In this context, the study seeks to answer the following research question: "How will water stress trends evolve in Türkiye until 2050, and what are the main socioeconomic, environmental, and sectoral factors determining this trend?" In this context, the study not only presents a new application to the scientific literature regarding the modeling of the multivariate structure of water stress but also produces insights that can guide Türkiye's sustainable water management policies.

LITERATURE REVIEW AND CONCEPTUAL BACKGROUND

Global and National Trends in Water Stress

Global water stress, resulting from the imbalance between increasing demand and limited renewable freshwater supply, has become an increasingly significant environmental and development problem. In global water balance studies, models that compare water demand and water availability at the river basin level on a monthly basis reveal that the proportion of people experiencing water scarcity is at least 40% higher than in traditional annual approaches (Wada et al., 2011). Similarly, analyses using the "blue water scarcity" indicator indicate that approximately two-thirds of the world's population experiences severe water scarcity for at least one month of the year, and approximately half a billion people experience persistent water scarcity throughout the year (Mekonnen and Hoekstra, 2016). These findings demonstrate that indicators that consider the geographical and temporal dimensions of water stress play a central role in sustainable water management discussions.

The global water scarcity literature conceptualizes water stress not only in terms of total abstraction but also in terms of environmental flow requirements, groundwater depletion, and water quality degradation. Hoekstra et al. (2012), in their approach comparing monthly bluewater footprints with bluewater availability, show that water scarcity is deepening, particularly in semiarid regions, due to the combination of agricultural irrigation expansion and climate change. The global water balance model developed by Wada, Van Beek, and Bierkens (2011) is another important reference point that allows quantitative monitoring of seasonal and regional variability in water stress.

In the context of Türkiye, a growing number of studies in recent years have documented both changes in precipitation patterns and regional drought trends in detail. Multi-method statistical analyses covering the period 1980–2019 indicate that precipitation across Türkiye exhibits significant spatial and temporal irregularities, with drought risk increasing particularly in summer (Aksu, 2021; Hinis and Geyikli, 2023; Sen et al., 2012). Studies using climatic drought indices such as SPEI reveal that drought intensity has gradually increased from western and central Anatolia to the southeast, and that several drought "hot spots" have emerged across the country (Şorman et al., 2018; Dalkilic, 2020, Pekpostalci et al., 2023). Meteorological, hydrological, and groundwater analyses conducted on the Konya Closed Basin indicate that groundwater drought has become particularly prevalent in the post-2008 period, and that withdrawals from agricultural irrigation have pushed the basin's water balance to an unsustainable level (Hinis and Geyikli, 2023; Dalkilic, 2020). Similar findings are observed in other closed or semi-closed basins, such as the semi-arid Burdur Basin, where increased irrigation water withdrawals are accompanied by shrinking lake areas, salinity, and ecosystem degradation (Çolak et al., 2022). This literature demonstrates that Türkiye's water security debate is based not only on total water potential but also on drought dynamics, groundwater management, and sectoral abstraction pressures at the basin scale; therefore, future water stress projections must take this multidimensional structure into account.

Sectoral Water Use Indicators and SDG Framework

Indicators used to assess water stress and water management performance have been redefined in recent years by the United Nations Sustainable Development Goals (SDG) framework. Two key global indicators stand out within SDG 6.4: 6.4.1, which measures water use efficiency, and 6.4.2, which measures the level of water stress (Biancalani and Marinelli, 2021; Giupponi et al., 2021). Developed by FAO and UN-Water, these indicators disaggregate water use by sector (agriculture, industry, services/municipality) and enable cross-country comparisons in terms of both water use per economic output and total pressure on renewable water resources (Vanham et al., 2018).

Water Use Efficiency (WUE) indicators measure the economic value created per unit of water, allowing for monitoring changes in the water intensity of growth. In their study evaluating SDG 6.4.1 and 6.4.2, Vanham et al. (2018) emphasize that obtaining WUE indicators, particularly across agriculture, industry, and services, is critical for linking water use to the "economic structure" (Giupponi et al., 2021; Cole et al., 2018). In the agricultural sector, due to the dominant share of blue water consumption by irrigated agriculture, indicators of irrigation water efficiency stand out as one of the most significant determinants of physical water scarcity. Mekonnen and Hoekstra (2016) show that agriculture accounts for approximately 70% of both water withdrawals and 90% of consumption globally, demonstrating that agricultural water withdrawals are central to water stress indicators, particularly in semiarid regions.

Industrial water use efficiency, on the other hand, reflects the capacity to create greater economic value by using less water in water-intensive processes such as manufacturing and energy production. Studies examining China, Europe, and large urban basins indicate that industry has become the most significant contributor to water stress in some regions, excluding agriculture, and that industrial water use exacerbates water scarcity in terms of both quantity and quality (pollution) (Becker et al., 2019; Liu et al., 2019; Zhang et al., 2024). WUE indicators calculated for the service/municipal sector are closely related to network loss/leakage rates, recycling capacity, and infrastructure investments in rapidly urbanizing regions. In cases where urban populations increase but service sector WUE remains low, water supply security becomes vulnerable (Mbavarira and Grimm, 2021; Chen et al., 2024).

The water stress indicator (SDG 6.4.2) defines total water withdrawal as the ratio of renewable freshwater resources remaining after deducting environmental flow requirements. In their study evaluating indicator 6.4.2, Vanham et al. (2018) argue that considering environmental flows and including additional components such as groundwater depletion and water quality are necessary for the indicator to be more meaningfully used in policymaking. Similarly, Marinelli et al. (2025) developed a new methodology for calculating indicator 6.4.2 at basin subregions and seasonal scales, highlighting the problem of "double counting" in water allocation, particularly between upstream and downstream subbasins. Studies specific to Türkiye indicate that indicators within the SDG 6.4 framework should be interpreted in conjunction with agricultural water withdrawals and agricultural WUE, particularly in closed basins. A recent study assessing agricultural water use efficiency in the Konya Closed Basin demonstrates that despite increasing agricultural productivity, there is significant potential for improvement in water use efficiency, and that the relationship between irrigation methods and water quality is a determining factor in water management. This framework reveals the analytical importance of using agricultural gravity, sectoral WUE and 6.4.2 water stress indicator together for studies discussing Türkiye's water future.

Economic and Climatic Drivers of Water Stress

The level and spatial distribution of water stress are determined not only by physical water availability but also by the combination of economic growth, structural transformations, urbanization dynamics, and climatic pressures. Global studies show that per capita GDP growth, industrial and service sector growth, and rising urbanization rates systematically increase water demand, particularly in developing countries (Wang et al., 2018; Zhao et al., 2021). Integrated water-energy security analyses in China reveal that rapid urbanization and industrial production create a strong coordination between water withdrawal and energy-related CO2 emissions, and that water scarcity is intertwined with energy security.

CO2 emissions and other greenhouse gas-driven climate change processes affect the water cycle, creating persistent changes in precipitation patterns, evaporation, and soil moisture, increasing the risks of both meteorological and hydrological droughts (Sen et al., 2012; Pekpostalci et al., 2023). The SPEI indicator, developed by Vicente-Serrano and colleagues, enables multi-scale monitoring of drought conditions based on the balance of precipitation and potential evaporation and is used in many studies analyzing the impact of climate change on drought dynamics. Studies conducted with SPEI and similar indices for Türkiye show that drought intensity has increased significantly in recent decades, especially in Central Anatolia, Southeastern Anatolia and Eastern Anatolia, and these regions have become "hot spots" in water stress (Pekpostalci et al., 2023).

Expanding irrigation areas for agricultural production, combined with economic growth and agricultural policies, further increase pressure on water resources, particularly groundwater. For example, in the semi-arid Burdur Basin, increased irrigation water withdrawals and climate change after 2004 have led to serious deteriorations in lake morphometry, salinity, and biotic communities (Çolak et al., 2022). In the Konya Closed Basin, increased well drilling and agricultural withdrawals have led to significant declines in groundwater levels and deepening hydrological drought, creating significant vulnerabilities for both water security and regional development.

A combined assessment of economic growth, CO2 emissions, urbanization, and drought indicators reveals that water security is determined not only by the amount of physical resources but also by socioeconomic structure and environmental governance capacity. Therefore, the combined use of macro indicators such as GDP, population, emissions, and drought indices in water stress models allows for the conceptualization of water as a political-economic issue, moving beyond a "natural scarcity" issue.

Forecasting Approaches in Water Demand and Stress Modelling

The water demand modeling literature offers a wide range of methods, ranging from simple polynomial regressions to advanced machine learning algorithms and hybrid models, for analyzing water use datasets with pronounced nonlinear relationships. While early studies largely relied on parametric approaches such as polynomial regression, various empirical analyses have demonstrated that these methods are limited in capturing the complex interactions of seasonality, sudden consumption spikes, and meteorological variables (Stańczyk et al., 2022). Therefore, the literature has shifted toward machine learning algorithms with more flexible functional structures, particularly for short-term and high-frequency water demand forecasting.

Random Forest (RF) has been found to be successful in identifying the most critical variables affecting water demand at the urban scale, demonstrating the decisive role of temperature, relative humidity, weekday/weekend effects, and time-of-day indicators on demand dynamics (Brentan et al., 2017). The advantage of RF is its ability to produce stable results, especially in short-term demand forecasts, due to its ability to model nonlinear relationships and high-order interactions between variables without parameter constraints. Studies comparing techniques such as RF, SVR, and Artificial Neural Networks have reported that RF provides high accuracy with complex data structures, while SVR achieves lower error rates in predicting demand peaks (Stańczyk et al., 2022).

Support vector regression (SVR) performs particularly well on short-term consumption series where the seasonal component is dominant. In some applications, SVR has been combined with Fourier series to create hybrid structures, and this integration has enabled more successful representation of periodic patterns in consumption (Brentan, Luvizotto Jr., et al., 2017). The literature also emphasizes that variable selection techniques using PCA, SOM, or RF-based importance measures make significant contributions to identifying the most informative indicators affecting demand and improving model performance (Brentan et al., 2017).

In recent years, the forecasting literature has shifted towards quantile regression and probabilistic forecasting approaches, which are more capable of modeling uncertainties in water demand. Studies on daily consumption data indicate that quantile regression algorithms exhibit superior performance, particularly in capturing uncertainty in extreme values (Papacharalampous & Langousis, 2022). In addition, combining denoising and subcomponent decomposition methods such as CEEMDAN used in data preprocessing with SVR, ANN, and GRU models has significantly increased short-term demand forecast accuracy (Hu et al., 2021).

Deep learning-based models are one of the fastest-growing areas in the water demand literature. Due to their ability to capture time-dependent consumption patterns, GRU and LSTM models produce lower error rates compared to classical methods in predicting hydrometeorological processes such as water demand and water quality. Li & Fu (2024) showed that LSTM and GRU models demonstrate superior performance in capturing nonlinear dependencies in consumption series and offer strong generalization capacity, especially for series with high volatility.

In general, the literature suggests that the success of water demand forecasting models depends on three key elements: (i) appropriately integrating meteorological indicators (temperature, humidity, precipitation), social time indicators (day/time effect), and past consumption into the model; (ii) using flexible methods that can capture nonlinear relationships; and (iii) accurately representing seasonality, sudden changes, and demand peaks. While polynomial regression still holds true in certain situations, it is clear that machine learning and deep learning-based hybrid models are increasingly becoming the dominant method set for water demand forecasting.

METHODOLOGY

This study aims to estimate Türkiye's water stress level by 2050. Thus, it aims to develop an empirical forecasting tool that can contribute to Türkiye's sustainable water management policies. In this context, a time-series-based forecasting model was created by combining environmental, socioeconomic, and sectoral indicators from past periods. This approach enables not only the analysis of historical trends but also the identification of potential future water stress scenarios (De Luna, 1998).

Figure 1. Data Analysis Process

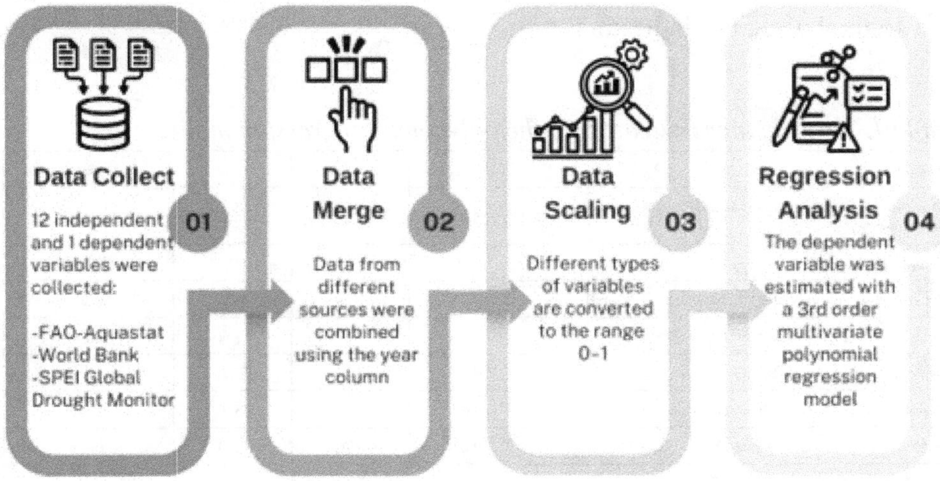

The dataset used in the analysis covers the period 1992–2022 and was compiled from three international sources. The first of these sources is the FAO Aquastat database, the primary reference for water use efficiency and water stress indicators. In this context, "Agricultural water withdrawal as a percentage of total renewable water resources (%)" indicators, SDG 6.4.1 indicators, and SDG 6.4.2 indicators were included in the model. Secondly, Türkiye's GDP (US$), CO_2 emissions (kt), and population data were obtained from the World Bank database. Thirdly, the SPEI Global Drought Monitor was used to obtain annual average Standard Precipitation-Evaporation Index (SPEI) values for Türkiye. This variable represents the impact of climatic drought conditions on water resources (George et al., 2016)

Data from different sources were matched on a yearly basis, and missing observations were removed to ensure continuity between 1992 and 2022. Since the measurement units of the variables are different, the min–max normalization method was applied to ensure statistical comparability (Chen et al., 2019). The model used in the study is based on the multivariate polynomial regression approach, which can capture non-linear relationships based on the assumption that water stress rates may not follow a linear trend. Polynomial models are widely used in short- and medium-term forecasting because they can flexibly explain trend components in time series (West & Harrison, 1989). Polynomial-based approaches provide more successful results than linear models in predicting long-term trends (De Luna, 1998; West & Harrison, 1989). In addition, to increase the predictive power of polynomial-based models, hybrid models using random forest and polynomial features have been shown to perform superiorly in air pollution predictions (Goyal & Goyal, 2024).

Polynomial regression models the nonlinear interactions between the dependent variable "Water Stress (%)" and time, economic indicators, environmental variables and sectoral water use indicators.

Table 1. Performance indicators of the polynomial regression model

variable	std	min	max	variance
Agricultural water withdrawal as % of total renewable water resources (%)	4.92	10.82	26.52	24.21
Industrial Water Use Efficiency (US$/m3)	77.53	23.06	260.73	6011.45
Irrigated Agriculture Water Use Efficiency (US$/m3)	0.08	0.14	0.40	0.01
Services Water Use Efficiency (US$/m3)	21.23	27.30	91.60	450.84
Water Use Efficiency (US$/m3)	2.47	7.72	17.79	6.13
Agricultural Sector Contribution to Water Stress (%)	7.89	17.01	43.38	62.30
Industrial Sector Contribution to Water Stress (%)	0.63	0.76	3.19	0.40
Municipal Sector Contribution to Water Stress (%)	0.69	3.10	5.69	0.47
Water Stress (%)	7.66	23.47	47.89	58.66
Annual SPEI Average (drought)	0.82	-2.10	1.05	0.67
GDP (Current US$)	3477.43	2240.63	12507.80	12092501.82
CO2 emissions (kt)	88710.70	139197.90	418098.20	7869587912.02
Population (total)	9071856.95	56302037.00	85341241.00	82298588605853.50

The model is expressed in general form as follows:

The model developed not only considers time-dependent trends but also the indirect effects of factors such as economic growth, population growth, industrial development, and climate variability on water stress. Similar methods have been successfully applied to predict environmental variables such as water demand and precipitation (Chen et al., 2019; George et al., 2016).

In this study, the model was trained using observations from the 1992–2022 period using a multivariate structure combining polynomial feature transformation (degree = 2) and Ridge regression. The forecast model was created using Ridge regularization. This method improves forecasting performance by reducing potential model instabilities such as multicollinearity and high dimensionality. The model then generated water stress projections for the 2023–2050 period, thus assessing potential water stress scenarios that Türkiye may face in the future.

Three key statistical metrics were used to evaluate the model's performance. The coefficient of determination (R^2) indicates how much of the variance in the dependent variable the model explains. The root mean square error (RMSE) measures the average deviation between the predicted and observed values (Chicco et al., 2021). Additionally, the variance inflation factor (VIF) values were examined to test the risk of multicollinearity among the independent variables. Model assumptions, such as normality of the error terms, homoscedasticity, and autocorrelation, were also tested using both graphical and statistical tests.

The validity of the model assumptions was measured using graphical and statistical tests. Analyses were conducted in the Python software environment. Pandas was used for data processing, Numpy for polynomial functions, Scikit-learn for model validation and error metrics, Matplotlib and Seaborn for visualization, and OpenPyXL for results export. The coding process was structured in accordance with the principle of repeatability.

Since the data were presented at an annual frequency, seasonal effects were not included in the model. Furthermore, the model is deterministic and based on current trends; therefore, potential changes in climate policies, new water management reforms, or unexpected environmental shocks were not directly reflected in the projections.

ANALYSIS AND RESULTS

This study presents the results of a multivariate polynomial regression model developed based on environmental, socioeconomic, and sectoral indicators for Türkiye for the period 1992–2022. In the first stage, annual data compiled from FAO Aquastat, the World Bank, and the SPEI Global Drought Monitor were matched, missing observations were removed, and min–max normalization was applied. In the second stage, descriptive statistics of the variables showed that water use efficiency, sectoral contribution to water stress, and macroenvironmental indicators had wide variance and trend ranges, methodologically supporting the preference of polynomial models that can capture nonlinear relationships (De Luna, 1998; West & Harrison, 1989). Third, the model was trained with polynomial feature transformation (degree = 2) and Ridge regularization to represent both trends and interactions among variables; this approach is a technique recommended in the literature for reducing the risk of multicollinearity (Chicco et al., 2021). In the final stage, model training and testing performance was evaluated, the total effects of the independent variables were calculated, and water stress projections were generated for the period 2023–2050. The results obtained throughout the analysis process are presented below.

Table 2. Predicted Water Stress Model Performance

Metric	Training set	Test set
R^2	0.9983	0.9835
RMSE	0.3004	0.7798

Model performance was evaluated using three key metrics: R^2, RMSE, and ridge regularization effect. Indicators for the training and test sets are presented in Table 3 below. High R^2 values indicate that the model explains more than 98% of the variance of the water stress variable. This level of performance is consistent with studies in the literature reporting successful representation of environmental time series with polynomial models (De Luna, 1998; George et al., 2016). The low RMSE on the test set indicates that the model avoids overfitting and has high generalization capacity. This is consistent with findings that ridge regularization successfully limits multicollinearity (Chicco et al., 2021). The low difference between training and test performance indicates that a model appropriate to the structure of the data set was established.

Table 3. Forecast Water Stress Model Performance

Metric	Value
R^2	0.964
RMSE	1.430

The values in Table 3 indicate that the multivariate polynomial model captures long-term trends with high accuracy. $R^2 = 0.964$ is similar to the performance levels of previous complex nonlinear models for water demand forecasting (Brentan, Meirelles, et al., 2017; Papacharalampous & Langousis, 2022)). The RMSE of 1.43 indicates that the variance in annual water stress is low and the projections are sufficiently reliable for policy analysis.

Figure 2. Forecasted Water Stress

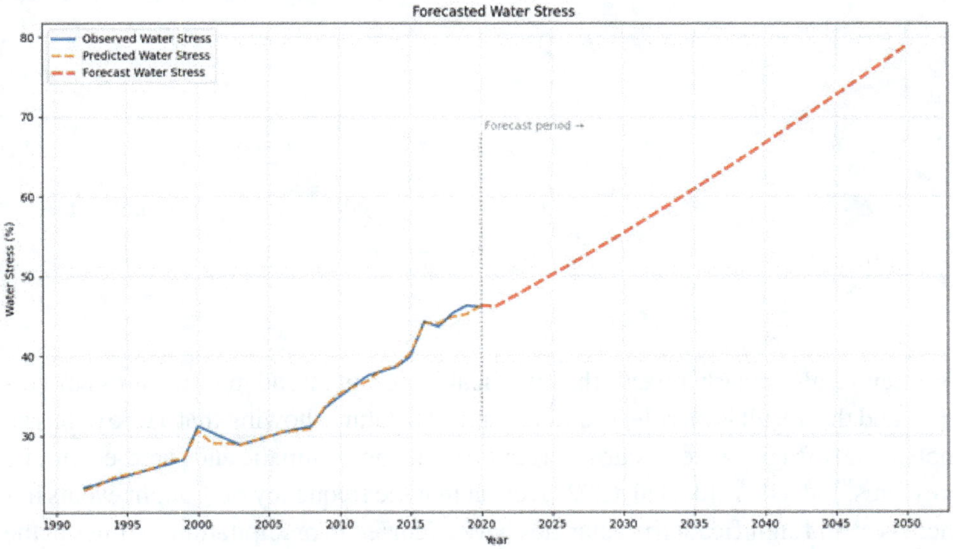

Graph 1 shows that the model largely accurately captured the observed water stress during the 1992–2022 period. The near correspondence between the observed values and the model predictions supports the strengths of the polynomial regression approach used in the study in representing nonlinear behavior in time series. This is consistent with the findings of West, West & Harrison (1989) who noted that polynomial models provide flexibility in explaining long-term trend components. It is also consistent with the findings of De Luna (1998), who demonstrated that polynomial structures offer more comprehensive interpretation possibilities in long-term forecasts compared to linear models. This robust capture of historical trends by the model demonstrates the advantages of polynomial-based approaches when working with high-variance variable sets in environmental time series. In the resulting data set, water withdrawal, economic indicators, CO2 emissions, and the drought index are observed to vary over wide ranges. The model's consistent representation of this multidimensional structure concretely confirms the suitability of this approach.

Figure 3. 2022-2050 Forecast water stress (%)

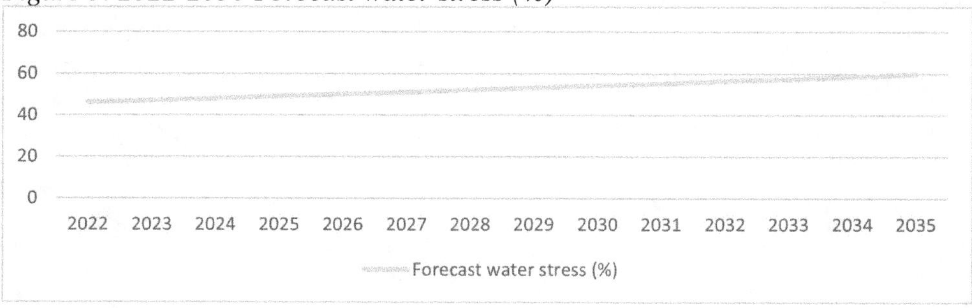

Graph 2 also clearly reveals the historical increasing trend in water stress in Türkiye, and this result is parallel to studies in the literature showing that Türkiye has an increasingly fragile water security structure under both climatic and socio-economic pressures. Indeed, Yetik et al. (2023) reveal that the frequency of drought events has increased and significant irregularities have occurred in precipitation regimes in the period after 1980. This finding is one of the main hydrometeorological mechanisms explaining the upward trend observed in the graph. Similarly, Çolak et al. (2022) indicate that the expansion of agricultural irrigation and the increase in industrial production in Türkiye have accelerated water demand, and these results support the acceleration observed in the period after 2000. Wada et al. (2011), who show that the increase in freshwater demand on a global scale drives the increase in water stress, and who emphasize that climate change has deepened water scarcity in the Mediterranean basin. Studies such as Chenoweth et al. (2011) also suggest that an upward trend in water stress is expected given Türkiye's geographical location. The 2023–2050 projection, starting at approximately 46% and rising steadily towards 2050, indicates that the potential water stress Türkiye will face in the future will intensify if current trends continue. This finding is consistent with the findings of Unnisa et al. (2023) who demonstrated that evapotranspiration in the Mediterranean basin has increased, making water supply more variable and fragile, and Vanham et al. (2018, 2021)), who reported that agricultural withdrawals significantly increase water stress in regions where environmental flow requirements cannot be met. Furthermore, the study by George et al. (2016) supports the notion that the declining SPEI over the years exacerbates drought conditions and increases uncertainty in the hydrological cycle. This suggests that the increase in water stress is also amplified by climate-based components.

Table 4. Variable Effects on Water Stress after Polynomial Transformation

Variable	Aggregated effect	Absolute magnitude	Effect
Population (total)	0.00000000849	0.0000002150	(+)
Agricultural Sector Contribution to Water Stress (%)	0.00000000502	0.0000000502	(+)
GDP (current US$)	0.00000000450	0.0000000454	(+)
CO_2 emissions (kt)	0.0000000431	0.0000000452	(+)
Agricultural water withdrawal (% TRWR)	0.00000000342	0.0000000342	(+)
Annual SPEI (drought index)	0.00000000323	0.0000000323	(+)
SDG 6.4.1 Water Use Efficiency (US$/m^3)	–0.00000000319	0.0000000319	(-)
Industrial Sector Contribution (%)	0.00000000287	0.0000000287	(+)
Services Water Use Efficiency (US$/m^3)	–0.00000000229	0.0000000229	(-)
Municipal Sector Contribution (%)	–0.00000000080	0.0000000080	(-)
Irrigated Agriculture WUE (US$/m^3)	–0.000000000207	0.000000000207	(-)
Industrial WUE (US$/m^3)	0.000000000169	0.000000000170	(+)

Table 4 shows the total effects of the variables in the model after polynomial transformation, and the signs indicate the direction of their contribution to water stress. The findings reveal that the population variable has the highest positive effect on water stress. Vanham et al. (2018) stated that population growth increases water stress by increasing both urbanization and municipal water demand. Similarly, Chenoweth et al. (2011) showed that population pressure structurally increases water demand, especially in the Mediterranean basin countries. Consistent with these findings, Table 4 shows that Türkiye's demographic growth is driving the water supply-demand balance into an increasingly fragile state. Consistent with this, the agricultural sector's contribution to water stress also emerges as a strong and positive determinant. Vanham et al. (2021) stated that agricultural water withdrawals systematically increase water stress in regions where environmental flow requirements cannot be met. UNESCO emphasized the dominant role of agricultural water withdrawals, showing that approximately 90% of total global consumption originates from agriculture. Indeed, the positive effect observed for this variable indicates the decisive role of the agricultural sector in Türkiye on water stress.

GDP, representing economic growth, and CO2 emissions, an indicator of industrial activity levels, also exhibit positive effects in the model. Mirziyoyeva & Salahodjaev (2023) found that increases in industrial production increase water stress through both water withdrawal and water consumption. Caporale et al. (2021) and Mendoza et al. (2021) showed that increased economic activity deepens the pressure on the climate system, strengthening evaporation and drought trends. Indeed, the results indicate that, while GDP and CO2 variables increase water stress, economic growth intensifies the pressure on Türkiye's water security.

The variables represented by negative effects in the model indicate factors that reduce water stress. The negative effects of total water use efficiency (WUE), service sector water use efficiency, and irrigated agriculture water use efficiency variables demonstrate the role of productivity in reducing water stress. Vanham et al. (2018) showed that increases in water use efficiency reduce sectoral water demand and thus contribute to the reduction of water stress. Similarly, they demonstrate that increasing WUE indicators have a mitigating effect on water stress. The negative sign representing municipalities' contribution to water stress can be attributed to the low variance in the normalized data set and is consistent with the findings of Vanham et al. (2021), who showed that municipal water use accounts for a limited share of total water withdrawal. Given the dominant position of agricultural and industrial sectors in total withdrawal, the relatively low and stable nature of municipal use is far from being a determinant of fluctuations in water stress, suggesting the structural cause of the negative effect.

Table 4 clearly demonstrates that demographic and economic pressures play a critical role in increasing water stress. Furthermore, the strong confirmation of the water stress-reducing effect of increases in water use efficiency aligns with the strategic importance emphasized by the studies of Vanham et al. (2018) and Vanham et al. (2021). Indeed, these results show that the water security problems that Türkiye will face in the future require not only sectoral but also economic and environmental interventions.

DISCUSSION AND CONCLUSION

The findings of this study indicate that Türkiye's water stress is exhibiting an increasing trend under the combined influence of both structural and climatic pressures. Polynomial model-based projections suggest that water stress will continue to increase towards 2050 and, if current trends continue, will exceed 60%. This trend supports the findings of Chenoweth et al. (2011) and Unnisa et al. (2023) that water scarcity in the Mediterranean basin is intensifying and hydrometeorological systems are becoming more fragile. Demographic pressures play the most significant role in increasing water stress. Population growth and urbanization are increasing municipal and service sector water demands, systematically weakening the water supply-demand balance (Vanham et al., 2018). This finding suggests that Türkiye's rapid urbanization will increase the burden on water management in the coming years. The positive contribution of the agricultural sector clearly reflects the structural nature of Türkiye's water stress. High agricultural withdrawal rates are becoming a key determinant of water stress, particularly in basins with limited environmental flow requirements (Vanham et al., 2021).

The positive impact of economic growth and CO2 emissions indicates that industrial activities increase water stress through both direct water demand and indirect climatic pressures. This impact demonstrates that the water-intensive nature of production structures poses a critical risk to sustainability (Csanádi et al., 2024). Furthermore, the positive contribution of the SPEI indicates that the increasing frequency of droughts in recent years has exacerbated water stress and that the impacts of climate change on water security are becoming increasingly evident (George et al., 2015; Yetik et al., 2023). Conversely, the negative sign of all WUE indicators suggests that water use efficiency offers strong remedial potential for Türkiye. Findings indicating that productivity increases reduce sectoral water demand (Vanham et al., 2018) directly align with the results of this study. Therefore, Türkiye's ability to alleviate water stress depends on prioritizing structural productivity increases as a strategic priority. These results clearly demonstrate that water stress is a multidimensional phenomenon shaped not only by climatic factors but also by demographic, economic, and sectoral processes. Therefore, solution approaches must incorporate integrated policies that consider this multilayered structure.

Agricultural water management is a priority policy area because it is the most dominant sectoral determinant of water stress. High water withdrawals in agriculture in Türkiye structurally exacerbate water stress, especially in narrow, closed basins. Therefore, accelerating irrigation modernization, expanding pressurized systems that reduce water losses, and restructuring basin-based crop patterns are among the most effective strategies to counter the upward trend in water stress. Developing mechanisms that encourage water efficiency in agricultural production will directly contribute to reducing the positive pressure on this sector.

Urban water management is another critical policy area for Türkiye, where demographic pressures are increasingly intensifying. Population growth and urbanization trends are rapidly increasing water demand, especially in metropolitan areas. Therefore, it is necessary to reduce loss and leakage rates, encourage the use of graywater in residential and commercial buildings, and strengthen urban infrastructure to support the use of recycled water. These interventions can help reduce municipal water demand to a more sustainable level in the long term.

Managing the interaction between water and carbon in industry is essential to mitigate the pressure that economic growth exerts on water stress. Transforming water-intensive production processes, standardizing wastewater recycling, and supporting the transition to low-carbon technologies offer the dual benefits of both reducing water demand and mitigating climatic pressures. Accelerating efficiency-focused transformation in industry has the potential to mitigate the cumulative effects of water stress.

Adapting to drought risk is a crucial policy area for Türkiye's water management capacity under climate change conditions. The positive contribution of the SPEI to water stress indicates that drought trends have intensified over the years. Therefore, strictly controlling groundwater use through quotas and monitoring systems, integrating wetland restoration into basin plans, and integrating drought early warning systems into decision-making mechanisms are critical to preserving the sustainability of water resources.

Finally, water use efficiency must be addressed as a strategic priority at the national level. The water stress-reducing impact of WUE indicators makes efficiency not merely a technical solution but also a policy tool that directly impacts economic and environmental sustainability. In this context, setting intersectoral efficiency targets, developing digital allocation systems, and implementing public incentives supporting the transition to water-efficient technologies can strengthen Türkiye's resilience capacity in water management.

Therefore, Türkiye's ability to move its water future toward a sustainable path depends on adopting a holistic water governance approach that not only increases water supply but also efficiently manages demand, supports sectoral transformation, and prioritizes climate adaptation. This study clearly outlines the areas where this transformation will have the strongest impact, providing a guiding framework for both scientific and policy-making processes.

REFERENCES

Aksu, H. H. (2021). Basin-based precipitation potential of Turkey. *Arabian Journal of Geosciences, 14*(23), 2470. Advance online publication. DOI: 10.1007/s12517-021-08841-2

Becker, D., Jungfer, C., & Track, T. (2019). Integrated industrial water management-challenges, solutions, and future priorities. *Chemieingenieurtechnik (Weinheim), 91*(10), 1367–1374. DOI: 10.1002/cite.201900086

Biancalani, R., & Marinelli, M. (2021). Assessing SDG indicator 6.4.2 'level of water stress' at major basins level. *UCL Open Environment, 3*(5). Advance online publication. DOI: 10.14324/111.444/ucloe.000026 PMID: 37228788

Brentan, B. M., Luvizotto, E.Jr, Herrera, M., Izquierdo, J., & Pérez-García, R. (2017). Hybrid regression model for near real-time urban water demand forecasting. *Journal of Computational and Applied Mathematics, 309*, 532–541. DOI: 10.1016/j.cam.2016.02.009

Brentan, B. M., Meirelles, G., Herrera, M., Luvizotto, E., & Izquierdo, J. (2017). Correlation Analysis of Water Demand and Predictive Variables for Short-Term Forecasting Models. Mathematical Problems in Engineering, 2017(1). DOI: 10.1155/2017/6343625

Caporale, G. M., Claudio-Quiroga, G., & Gil-Alana, L. A. (2021). Analysing the relationship between CO2 emissions and GDP in China: A fractional integration and cointegration approach. *Journal of Innovation and Entrepreneurship, 10*(1), 32. DOI: 10.1186/s13731-021-00173-5

Chen, G., Long, T., Bai, Y., & Zhang, J. (2019). A Forecasting Framework Based on Kalman Filter Integrated Multivariate Local Polynomial Regression: Application to Urban Water Demand. *Neural Processing Letters, 50*(1), 497–513. DOI: 10.1007/s11063-019-10001-3

Chen, L., Chen, Z., Liu, Y., Lichfouse, E., Jiang, Y., Hua, J., Osman, A. I., Farghali, M., Huang, L., Zhang, Y., Rooney, D. W., & Yap, P. S. (2024). Benefits and limitations of recycled water systems in the building sector: A review. *Environmental Chemistry Letters, 22*(2), 785–814. DOI: 10.1007/s10311-023-01683-2

Chenoweth, J., Hadjinicolaou, P., Bruggeman, A., Lelieveld, J., Levin, Z., Lange, M. A., Xoplaki, E., & Hadjikakou, M. (2011). Impact of climate change on the water resources of the eastern Mediterranean and Middle East region: Modeled 21st century changes and implications. *Water Resources Research, 47*(6), 2010WR010269. Advance online publication. DOI: 10.1029/2010WR010269

Chicco, D., Warrens, M. J., & Jurman, G. (2021). The coefficient of determination R-squared is more informative than SMAPE, MAE, MAPE, MSE and RMSE in regression analysis evaluation. *PeerJ. Computer Science*, *7*, e623. DOI: 10.7717/peerj-cs.623 PMID: 34307865

Çolak, M. A., Öztaş, B., Özgencil, İ. K., Soyluer, M., Korkmaz, M., Ramírez-García, A., Metin, M., Yılmaz, G., Ertuğrul, S., Tavşanoğlu, Ü. N., Amorim, C. A., Özen, C., Apaydın Yağcı, M., Yağcı, A., Pacheco, J. P., Özkan, K., Beklioğlu, M., Jeppesen, E., & Akyürek, Z. (2022). Increased Water Abstraction and Climate Change Have Substantial Effect on Morphometry, Salinity, and Biotic Communities in Lakes: Examples from the Semi-Arid Burdur Basin (Turkey). *Water (Basel)*, *14*(8), 1241. DOI: 10.3390/w14081241

Cole, M. J., Bailey, R. M., Cullis, J. D. S., & New, M. G. (2018). Water for sustainable development in the Berg Water Management Area, South Africa. *South African Journal of Science*, *114*(3/4), 10. Advance online publication. DOI: 10.17159/sajs.2018/20170134

Dalkilic, H. Y., Bayçinar, M., & Samui, P. (2020). Evaluation of combined use of drought indices in the case of Konya closed basin. *International Journal of Global Warming*, *23*(2), 169–190. DOI: 10.1504/IJGW.2021.112895

De Luna, X. (1998). Projected polynomial autoregression for prediction of stationary time series. *Journal of Applied Statistics*, *25*(6), 763–775. DOI: 10.1080/02664769822756

George, J., Janaki, L., & Parameswaran Gomathy, J. (2016). Statistical Downscaling Using Local Polynomial Regression for Rainfall Predictions – A Case Study. *Water Resources Management*, *30*(1), 183–193. DOI: 10.1007/s11269-015-1154-0

Giupponi, C., Gain, A. K., & Farinosi, F. (2021). Spatial assessment of water use efficiency (SDG indicator 6.4.1) for regional policy support. *Frontiers in Environmental Science*, *6*, 141. Advance online publication. DOI: 10.3389/fenvs.2018.00141

Goyal, K., & Goyal, S. (2024). Predicting PM2.5 Air Quality Using Random Forest Regression Enhanced with Polynomial Features. 2024 IEEE 12th International Conference on Intelligent Systems (IS), 1–6. DOI: 10.1109/IS61756.2024.10705219

Hinis, M. A., & Geyikli, M. S. (2023). Accuracy evaluation of standardized precipitation index (SPI) estimation under conventional assumption in Yeşilırmak, Kızılırmak, and Konya closed basins, Turkey. *Advances in Meteorology*, *2023*, 1–13. Advance online publication. DOI: 10.1155/2023/5142965

Hoekstra, A. Y., Mekonnen, M. M., Chapagain, A. K., Mathews, R. E., & Richter, B. D. (2012). Global monthly water scarcity: Blue water footprints versus blue water availability. *PLoS One*, *7*(2), e32688. Advance online publication. DOI: 10.1371/journal.pone.0032688 PMID: 22393438

Hu, S., Gao, J., Zhong, D., Deng, L., Ou, C., & Xin, P. (2021). An Innovative Hourly Water Demand Forecasting Preprocessing Framework with Local Outlier Correction and Adaptive Decomposition Techniques. *Water (Basel)*, *13*(5), 582. Advance online publication. DOI: 10.3390/w13050582

Li, D., & Fu, G. (2024). Deep Learning Model-Based Demand Forecasting for Secondary Water Supply in Residential Communities: A Case Study of Shanghai City, China. IEEE Access. DOI: 10.1109/ACCESS.2023.3288817

Liu, J., Li, X., Yang, H., Han, G., Liu, J., Zheng, C., & Zheng, Y. (2019). The water-energy nexus of megacities extends beyond geographic boundaries: A case of Beijing. *Environmental Engineering Science*, *36*(7), 778–788. Advance online publication. DOI: 10.1089/ees.2018.0553 PMID: 31346304

Marinelli, M., Biancalani, R., Joyce, B., & Djihounessi, M. B. (2025). A New Methodology to Estimate the Level of Water Stress (SDG 6.4.2) by Season and by Sub-Basin Avoiding the Double Counting of Water Resources. *Water (Basel)*, *17*(10), 1543. Advance online publication. DOI: 10.3390/w17101543

Mbavarira, T. M., & Grimm, C. (2021). A systemic view on circular economy in the water industry: Learnings from a Belgian and Dutch case. *Sustainability (Basel)*, *13*(6), 3313. Advance online publication. DOI: 10.3390/su13063313

Mekonnen, M. M., & Hoekstra, A. Y. (2016). Four billion people facing severe water scarcity. *Sustainability*, *2*(2), e1500323. Advance online publication. DOI: 10.1126/sciadv.1500323 PMID: 26933676

Mendoza, A. A., Garret, T., Reyes, K. D. D., Soriano, P. A., & Cabauatan, R. (2021). The Impact of CO2 Emissions on the GDP per Capita, Employment Rate and Energy Consumption of China, Korea and Japan. [MJSSH]. *Malaysian Journal of Social Sciences and Humanities*, *6*(11), 315–333. DOI: 10.47405/mjssh.v6i11.1134

Mirziyoyeva, Z., & Salahodjaev, R. (2023). Renewable energy, GDP and CO2 emissions in high-globalized countries. *Frontiers in Energy Research*, *11*, 1123269. Advance online publication. DOI: 10.3389/fenrg.2023.1123269

Papacharalampous, G., & Langousis, A. (2022). Probabilistic Water Demand Forecasting Using Quantile Regression Algorithms. *Water Resources Research*, *58*(6), e2021WR030216. Advance online publication. DOI: 10.1029/2021WR030216

Pekpostalci, D. S., Tur, R., Mehr, A. D., Ghaffari, M. A. V., Dabrowska, D., & Nourani, V. (2023). Drought Monitoring and Forecasting across Turkey: A Contemporary Review. *Sustainability (Basel)*, *13*(7), 6080. Advance online publication. DOI: 10.3390/su15076080

Sen, B., Topcu, S., Türkeş, M., & Warner, J. (2012). Projecting climate change, drought conditions and crop productivity in Turkey. *Climate Research*, *52*(1), 175–191. DOI: 10.3354/cr01074

Şorman, A. Ü., Mehr, A. D., & Hadi, S. J. (2018). Study on spatial-temporal variations of Meteorological-Agricultural droughts in Turkey. The International Archives of the Photogrammetry, Remote Sensing and Spatial Information Sciences, XLII(3/W4). DOI: 10.5194/isprs-archives-XLII-3-W4-483-2018

Stańczyk, J., Kajewska-Szkudlarek, J., Lipinski, P., & Rychlikowski, P. (2022). Improving short-term water demand forecasting using evolutionary algorithms. *Scientific Reports*, *12*(1), 13522. Advance online publication. DOI: 10.1038/s41598-022-17177-0 PMID: 35941276

Unnisa, Z., Govind, A., Lasserre, B., & Marchetti, M. (2023). Water Balance Trends along Climatic Variations in the Mediterranean Basin over the Past Decades. *Water (Basel)*, *15*(10), 1889. DOI: 10.3390/w15101889

Vanham, D., Alfieri, L., Flörke, M., Grimaldi, S., Lorini, V., de Roo, A., & Feyen, L. (2021). The number of people exposed to water stress in relation to how much water is reserved for the environment: A global modelling study. *The Lancet. Planetary Health*, *5*(11), e766–e774. DOI: 10.1016/S2542-5196(21)00234-5 PMID: 34774120

Vanham, D., Hoekstra, A. Y., Wada, Y., Bouraoui, F., de Roo, A., Mekonnen, M. M., van de Bund, W. J., Batelaan, O., Pavelic, P., Bastiaanssen, W. G. M., Kummu, M., Rockström, J., Liu, J., Bisselink, B., Ronco, P., Pistocchi, A., & Bidoglio, G. (2018). Physical water scarcity metrics for monitoring progress towards SDG target 6.4: An evaluation of indicator 6.4.2 "Level of water stress.". *The Science of the Total Environment*, *613–614*, 218–232. DOI: 10.1016/j.scitotenv.2017.09.056 PMID: 28915458

Wada, Y., van Beek, L. P. H., & Bierkens, M. F. P. (2011). Modelling global water stress of the recent past: On the relative importance of trends in water demand and climate variability. *Hydrology and Earth System Sciences*, *15*(12), 3785–3808. DOI: 10.5194/hess-15-3785-2011

Wang, Y., Xiao, W., Wang, Y., Hou, B., Yang, H., Zhang, X., Yang, M., & Zhu, L. (2018). Exploring City Development Modes under the Dual Control of Water Resources and Energy-Related CO2 Emissions: The Case of Beijing, China. *Sustainability (Basel), 10*(9), 3155. Advance online publication. DOI: 10.3390/su10093155

West, M., & Harrison, J. (1989). Polynomial Trend Models (pp. 201–228). Yetik, A. K., Arslan, B., & Şen, B. (2023). Trends and Variability in Precipitation Across Turkey: A Multi-Method Statistical Analysis. https://doi.org/DOI: 10.1007/978-1-4757-9365-9_7

Zhang, Z., Shan, Y., Zhao, D., Tillotson, M. R., Cai, B., Li, X., Zheng, H., Zhao, C., Guan, D., Liu, J., & Hao, Y. (2024). City level water withdrawal and scarcity accounts of China. *Scientific Data, 11*(449), 449. Advance online publication. DOI: 10.1038/s41597-024-03115-4 PMID: 38702307

Zhao, D., Liu, J., Sun, L., Ye, B., Hubacek, K., Feng, K., & Varis, O. (2021). Quantifying economic-social-environmental trade-offs and synergies of water-supply constraints. *Water Research, 195*, 116986. Advance online publication. DOI: 10.1016/j.watres.2021.116986 PMID: 33721677

Chapter 9

The Relationship Between Climate Change and Marine Economics in Türkiye

Bilal Göde
https://orcid.org/0000-0001-8377-5909
Pamukkale University, Turkey

ABSTRACT

Climate change is a threat affecting every aspect of life. The scale and impact of this threat must be determined, and measures must be taken. Türkiye is located in one of the regions that will be most affected by climate change. This study examines the effects of climate change on Türkiye's marine economy. Using data from 2000-2023, the relationship between sea surface temperature (stt), acidity level (pH), exchange rate, and oil price, and their contributions to the marine economy, was investigated using the ARDL cointegration test, and it was concluded that there is cointegration between the variables. While global warming is thought to have positive short-term effects on tourism, in the long term, extreme temperature increases and the risk of drought may lead to negative effects. Global climate change is also expected to have negative impacts on the fishing sector.

INTRODUCTION

By the beginning of the 21st century, there had been a significant shift in global priorities, and as a result of increased environmental awareness, global warming rose to the top of the list of priorities. The impact of human activity on global warming

DOI: 10.4018/979-8-3373-1777-9.ch009

became a major topic of debate (Alper et al., 2024). Policies implemented after World War II have led to negative consequences for the environment. The exploitation of the environment and natural resources as raw materials during industrialization and the neglect of the environment for the sake of economic growth are among the major causes of environmental degradation (Yavuz and Ergen, 2022: p.114). Climate change-related challenges are more severe in developing countries than in developed countries, requiring the implementation of innovative and effective environmental policies. The increase in carbon emissions also deepens and exacerbates the effects of climate change (Köktaş et al. 2024: 365).

Increasing greenhouse gas emissions since the Industrial Revolution have led to a steady rise in global temperatures and irreversible changes in the climate system. One of the areas most affected by these changes is the oceans and seas. Rising sea surface temperatures, ocean acidification, sea level rise, and declining oxygen levels directly threaten marine ecosystems and the economic sectors dependent on them.

The concept of the "blue economy" refers to an economic model that focuses on the sustainable use of oceans, seas, and coastal resources. Türkiye is a country with high blue economy potential due to its long coastline and rich marine resources, surrounded by seas on three sides. Tourism, fishing, maritime transport, and coastal investments are of vital importance to the Turkish economy. However, the pressures that climate change exerts on these sectors put the country's development goals and ecological sustainability at risk.

The aim of this study is to assess the current and projected impacts of climate change on Türkiye's blue economy, particularly on the tourism and fishing sectors, from an interdisciplinary perspective. The study examines the causal relationships between climate variables (SST, pH) and blue economy performance using climate projections, sectoral statistics, and econometric modeling methods. Furthermore, trends in tourism demand and fisheries production were analyzed to reveal the climate vulnerabilities of these sectors. Based on these comprehensive analyses, the study aimed to develop policy recommendations to make Türkiye's blue economy resilient to the adverse effects of climate change.

CLIMATE CHANGE AND GLOBAL WARMING

Climate change refers to permanent changes in the climate caused by natural factors as well as, in particular, human factors. These human-induced effects are generally negative, i.e., disruptive in nature. Global warming refers to permanent increases in average surface temperatures, both globally and regionally. Global warming is seen as one of the components within the concept of climate change. Climate change and global warming are among the most important issues of the coming century for humanity and other living beings (Yavuz et al., 2024: 67260). Global warming and climate change are related but distinct concepts. Global warming is one of the causes of climate change. Global warming can lead to factors such as melting glaciers and rising sea and ocean levels. Climate change also encompasses other issues such as changes in seasons and precipitation patterns (Climate Change Presidency, Basic Concepts).

The basic elements of the climate system are the atmosphere, oceans, ice-glacier cover, land, and biosphere. This system as a whole can be affected by both external factors and natural processes, and partial effects can also be seen. The world has experienced many climate change processes over a period of 4.5 billion years. Human activities, which accelerated particularly after the Industrial Revolution, especially since the 1750s, have caused a significant increase in carbon emissions and changes in the composition of the atmosphere. The accumulation of CO_2, the most important greenhouse gas in the atmosphere, has risen from 280 ppm before industrialization to 425 ppm in August 2025. The area of Arctic ice, which was around 7 million km^2 in 1979, has decreased to 4.3 million km^2 in 2024. Global sea levels have risen by approximately 91 mm from 1993 to the present (July 2025).

Figure 1. Climate Change Projections

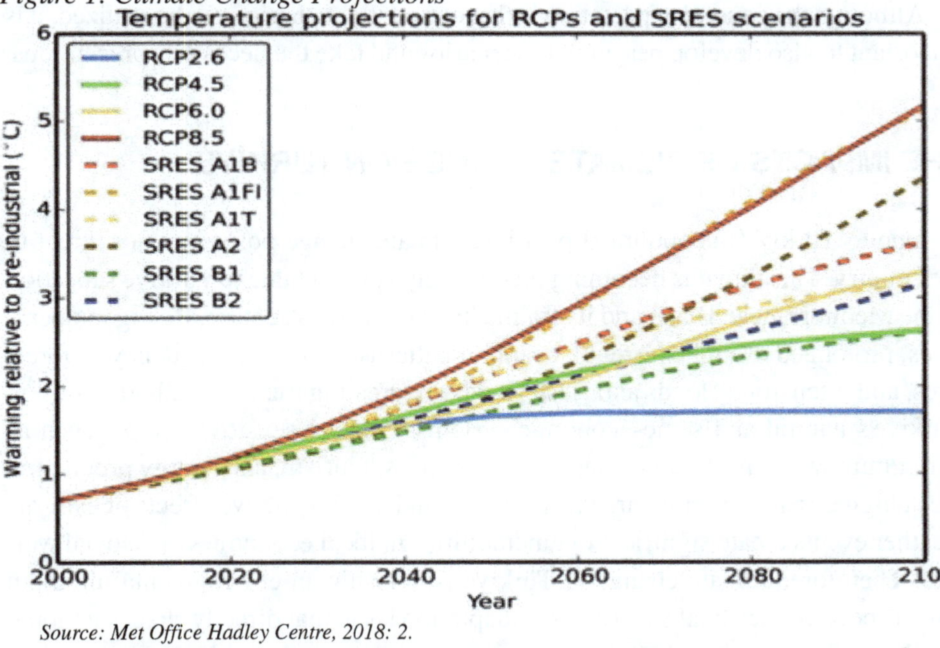

Source: Met Office Hadley Centre, 2018: 2.

**Global mean temperature projections from a climate model (called MAGICC6) relative to a pre-industrial average (1850-1900) for RCP2.6 (blue), RCP4.5 (green), RCP6.0 (yellow) and RCP8.5 (red) and the older SRES scenarios (dashed colored lines).*

Predicting future climate changes and climate patterns is crucial for all countries in terms of future planning. To this end, these predictions are made within the framework of various scenarios. Scenarios are created based on various expectations. In order to consider all possibilities, pessimistic scenarios are created alongside optimistic ones.

RCP scenarios have four main projections at various levels. The most important difference in distinguishing RCP scenarios is the average global surface temperature increase predicted by 2100.

- RCP2.6 has the potential to limit warming to between 1.5°C and 2°C (compared to pre-industrial levels). It is also the most optimistic scenario compared to other models.
- RCP4.5 and RCP6.0 assume moderate warming, projecting an increase limited to approximately 2°C to 4°C. These are moderately optimistic scenarios.
- RCP8.5 is the most pessimistic scenario regarding future expectations. It is the scenario with the most severe climate change impacts, with average temperature increases exceeding 4°C and even approaching 5°C in some models (IPCC, 2014:8).

Although the general expectation is that optimistic scenarios can be realized, it is important to also develop pessimistic scenarios and take the necessary precautions.

THE IMPACTS OF CLIMATE CHANGE ON TÜRKİYE

Due to Türkiye's geopolitical position, climate change poses a major threat to the country. This threat is becoming increasingly apparent due to Türkiye's location in the Mediterranean Basin and its fragile hydroclimatic structure. Rising temperatures, prolonged droughts, irregular rainfall patterns, increased frequency of forest fires, and intensified floods and flash floods are creating multifaceted pressures on Türkiye's natural and socio-economic systems. Climate-sensitive sectors such as agriculture, water resource management, fisheries, tourism, and energy production are at higher risk; in particular, water scarcity and the destructive effects of extreme weather events create significant vulnerabilities in local economies and social welfare. Therefore, climate change in Türkiye is not only an environmental problem but has become a critical security and adaptation issue that directly shapes national development strategies.

Figure 2. Distribution of Türkiye's 2024 Monthly Average Temperature Values Compared to the 2023 and 1970-2024 Monthly Average Temperatures

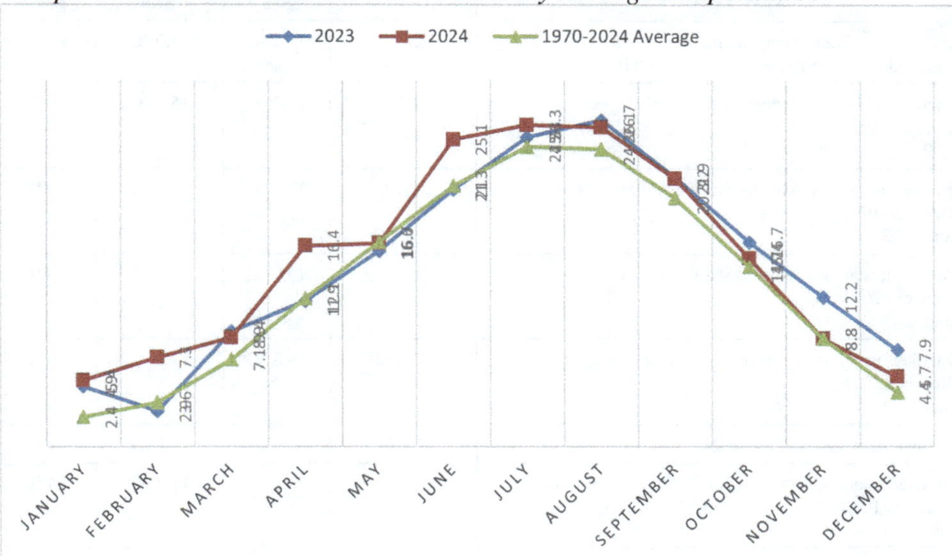

Source: General Directorate of Meteorology, Ministry of Environment, Urbanization, and Climate Change, 2025.

The figure shows that Türkiye is experiencing a clear warming trend. Both the 2023 and 2024 temperature data consistently exceed the 1970-2024 average, with some exceptions. This indicates that the likelihood of these temperature changes being permanent is strong. The most critical deviations in the figure were observed during the summer months, when temperatures were highest, and in September and October, which could be considered shoulder months. This confirms that the Mediterranean basin, which includes Türkiye, is one of the regions most affected by global warming. In terms of its effects, it causes major problems such as severe droughts, sea level rise, and ocean acidification (UFM Secretariat, 2019:5).

Table 1. Average Sea Water Temperatures in Türkiye

	Mediterranean	Aegean Sea	Marmara Sea	Black Sea
Average Sea Water Temperature in the Mediterranean Between 1970 and 1980	21°C	18.2 °C	15.0 °C	15.1 °C
Mediterranean Sea Average Sea Water Temperature Between 1981- 1991	21.1°C	18.3 °C	15.2 °C	14.8 °C
Mediterranean Sea Average Sea Water Temperature Between 1992 and 2002:	21.4°C	18.5 °C	15.4 °C	14.9 °C
Mediterranean Sea Average Sea Water Temperature Between 2003 and 2013	21.9 °C	19.0 °C	15.8 °C	15.6 °C
Mediterranean Sea Average Sea Surface Temperature Between 2014 and 2024	22.3°C	19.8 °C	17.2 °C	16.5 °C
Change Between 1970 and 2024	1.3°C	1.6°C	2.2°C	1.4°C
Mediterranean Sea Average Sea Surface Temperature Between 1970 and 2024	21.5°C	18.8 °C	15.7 °C	15.4 °C

Source: General Directorate of Meteorology, 2025.

Climate change is one of the biggest and most important threats we face globally. This threat is particularly noticeable in the Mediterranean region. The Mediterranean basin is warming at a rate 20% higher than the global average. This situation will bring serious problems to the basin in both the short and long term. Sea levels are expected to rise by 1 meter by 2100, affecting one-third of the region's population (WWF France, 2021: 3).

Looking at sea water temperatures over the period 1970-2024, a steady increase is evident. Mediterranean water temperature has risen from 21°C to 22.3°C. Aegean Sea water temperature has risen from 18.2°C to 19.8°C. The water temperature of the Marmara Sea has risen from 15°C to 17.2°C. The water temperature of the Black Sea has risen from 15.1°C to 16.5°C. These increases are also significantly affecting the marine ecosystem.

Figure 3. Türkiye Temperature Anomaly Projections

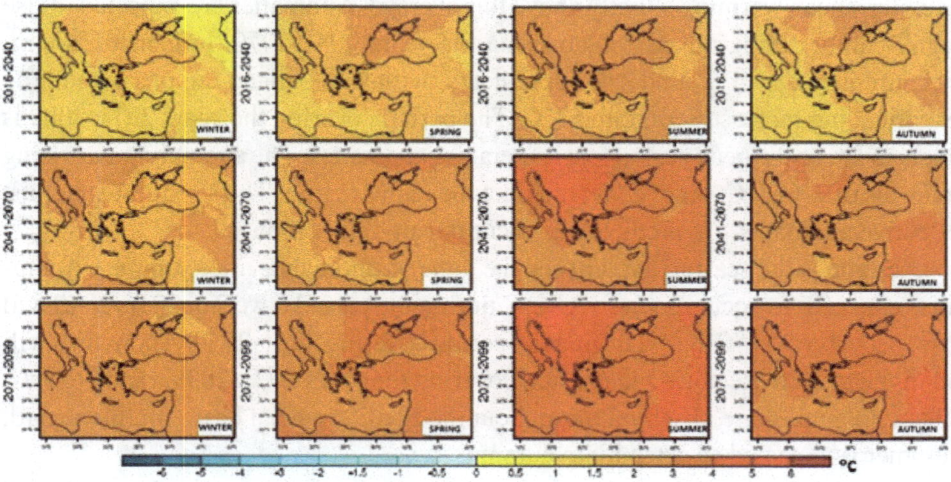

Source: Demircan et al. 2017: 6.

Figure 3 contains temperature increase projections for the Mediterranean basin specific to Türkiye for the periods 2016-2099. Separate projections have been created for each season for each period. On the color scale, darker colors represent higher temperature increases. This scenario covers the periods 2016- 2040, 2041-2070, and 2071-2099. HadGEM2-ES was used based on the RCP4.5 scenario.

Looking specifically at the winter season, anomalies that could be considered high are not frequently observed in the period covering 2016-2040, but the trend is positive. In the period covering 2041-2070, however, anomalies rise to more serious levels. Anomalies of around 2°C are predicted. It is possible to see that Türkiye and the eastern Mediterranean are significantly warmer. In the period covering the years 2071-2099, temperature anomalies exceed 3°C. Anomalies exceeding 3.5°C are expected in the eastern Mediterranean. Despite being winter months, cold anomalies are predicted to disappear.

Spring will be a season with slightly more anomalies compared to winter. Anomalies are projected to generally range between 1°C and 2°C during the period covering the years 2016-2040. Anomalies are expected to be slightly higher in the Aegean and Mediterranean regions. It is estimated that warming will rise to the range of 2.5°C-3°C in the period covering the years 2041-2070, and anomalies are expected to be distinctly felt. It is thought that anomalies will increase to around 3°C-4°C during the period covering the years 2071-2099, making them more noticeable. Anomalies are predicted to be seen across the entire region.

Summer is expected to be the season when temperature anomalies will be most intense and widespread. During the period covering the years 2016-2040, warming is expected to be between approximately 1.5°C and 2.5°C, with the eastern Mediterranean warming slightly above this average. Anomalies are expected to rise significantly during the period covering the years 2041-2070. In general, a temperature increase of around 3°C is predicted, with some inland regions potentially experiencing increases of around 4°C. The period covering the years 2071-2099 is expected to be the peak period for temperature anomalies. The anomaly wave, rising to levels of 4°C-6°C, is expected to affect the entire region. The strongest and most effective anomalies are predicted to occur during this season.

Autumn is the season that will see the greatest anomalies after summer. Anomalies expected to occur around 1°C-2°C during the period covering 2016-2040 will rise to levels of 2.5°C-3.5°C during the period covering 2041-2070. Anomalies will continue to increase during the period covering the years 2071-2099, rising to 4°C. These anomalies are observed throughout the Mediterranean basin. In this regard, summer is expected to extend into autumn.

In general, temperature anomalies are observed to increase in all periods and seasons. Warming is more pronounced in the fall and summer months. Although increases are also observed in the spring and winter seasons, they remain below the levels seen in the fall and summer seasons. The eastern Mediterranean and Levant[1] coastlines (southern Türkiye, Syria, Lebanon, Israel, northern Egypt) are expected to experience the highest warming, particularly in the period 2071-2099. These high anomalies expected to occur in the period 2071-2099 are predicted to have significant effects on both the sea and the land.

Looking at the countries included in the projection, it is predicted that these anomalies will not be limited to Türkiye, but will affect all countries bordering the Mediterranean, particularly the eastern Mediterranean countries. Considering that the RCP 4.5 scenario is a relatively optimistic one, the expectation that even projections based on these scenarios could show a 6°C anomaly highlights the importance of the issue. This further emphasizes the importance of taking necessary measures. These projections particularly support the view that the Mediterranean basin is a hotspot.

The anomalies expected to occur according to the projections will have significant effects on both marine and terrestrial ecosystems. By the end of the century, serious socio-economic changes are expected to occur in the Mediterranean region.

Figure 4. Türkiye Rainfall Anomaly Projections

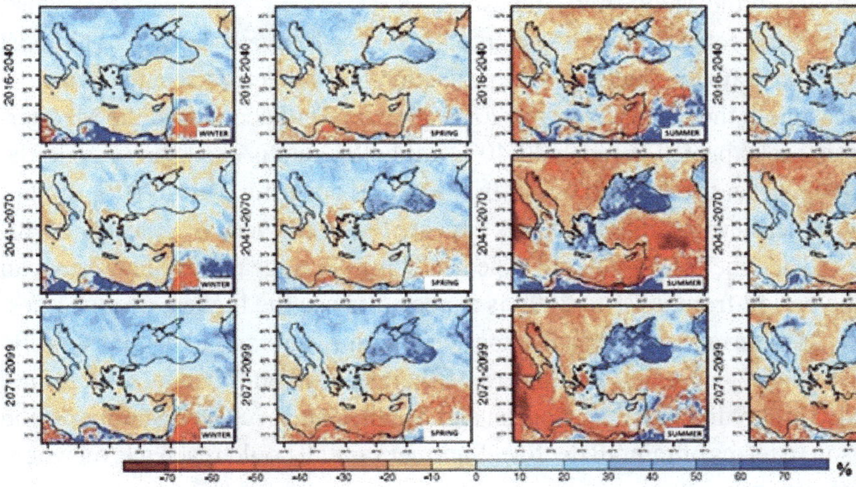

Source: Demircan et al. 2017: 6.

Figure 4 contains temperature increase projections for the Mediterranean basin, specifically for Türkiye, for the periods 2016-2099. Separate projections have been created for each season for each period. The shift of colors in the projection towards red tones indicates precipitation decreases, while the shift towards blue tones indicates precipitation increases. Looking at all seasons and periods, it clearly shows that the Mediterranean basin is heading towards a serious drought risk, especially in the summer and autumn seasons.

Looking specifically at the winter season, there are slightly positive anomalies in the western parts of Türkiye during the period covering 2016-2040. In the eastern Mediterranean region, insufficient increases ranging from neutral to 10-20% are observed. In the period covering 2041-2070, slight increases can be observed in western Türkiye, while serious decreases of up to 40% are expected in the southern Mediterranean basin. In the period covering 2071-2099, although slight increases may be observed throughout Türkiye, decreases are noticeable in the Mediterranean basin. In general, the winter season appears to be the least affected season in terms of precipitation due to its stable projection.

Although spring has more anomalies than winter, it can be said to have a medium level of variability compared to other seasons. No extreme changes are observed across Türkiye during the period covering 2016-2040. Rainfall losses are noticeable during the period covering 2041-2070. Although there are some complex structures in the Mediterranean region, the overall trend is negative, and rainfall losses are observed. Rainfall losses peak in the period covering the years 2071-2099. Rainfall losses, which are expected to occur at levels of 20-40%, could have a particularly serious impact on the Mediterranean basin.

Summer is expected to be the season with the most severe drought. The most pronounced and dramatic changes in the projections are expected to occur during the summer season. In the period covering 2016-2040, rainfall declines in the Mediterranean basin are expected to reach 20-40%, while this rate is expected to be in the range of 10-20% on the Turkish coast. In the period covering 2041- 2070, rainfall decline in the eastern Mediterranean is expected to reach very high levels of 40-60%. Rainfall declines of 20-40% are expected along the Aegean and Mediterranean coasts. The period from 2071 to 2099 is expected to see the highest decline rates. Most of the Mediterranean basin will experience significant rainfall declines. While declines of 60-80% are expected in most of the basin, rainfall declines exceeding 40% are also highly likely on the Aegean and Mediterranean coasts. Summer is the worst season in terms of rainfall decline. Summer rainfall could reach a level where it is almost non-existent.

After summer, the season with the most significant rainfall decline is autumn. During the period covering 2016-2040, rainfall declines of around 20% are expected in Türkiye and the eastern Mediterranean. Although slight increases appear to be occurring in the western Mediterranean, the general trend is negative. A strong aridification is expected across the entire region for the period covering the years 2041-2070. The expected rainfall decline of 20%-40% in Türkiye rises to 40%-60% levels, especially in the eastern Mediterranean. The most serious rainfall declines are expected to be seen for the period covering the years 2071-2099. Precipitation declines of around 30-50% are likely for Türkiye. Given these levels, the autumn season following the summer season is expected to be a period when drought will be severely felt.

If we make a general assessment by period, rainfall in the winter season is expected to increase slightly in the Aegean coast, Eastern Black Sea, and Eastern Anatolia during the period covering 2016- 2040, while spring rainfall is expected to decrease by around 20% in most of the country, except for the Aegean coast and eastern Anatolia. For the period covering 2041-2070, winter rainfall is expected to decrease by around 20% in eastern and southeastern Anatolia and the central and eastern parts of the Mediterranean region. In the eastern Anatolia region, where summer rainfall is particularly important, summer rainfall is also predicted to decrease by around 30%. In autumn rainfall, it is estimated that there will be a decrease in rainfall throughout the country, except for the coastal Aegean and a small part of inner Anatolia. In the period covering 2071-2099, autumn rainfall is expected to decrease throughout the country, while winter rainfall is expected to decrease by 10%, especially in coastal areas. Summer rainfall will decrease critically, reaching levels of 40% across the country, except for the Aegean, Marmara, and Black Sea coasts.

When rainfall and temperature projections are considered together, the expected decrease in rainfall and increase in temperatures will bring significant changes for all of Türkiye, especially the Mediterranean region. Although there may be some advantages in certain respects, largely negative effects are expected. Increasing population and economic growth increase water demand, while the downward trend in rainfall raises the issue of water sufficiency as a major problem for policymakers.

Figure 5. Amount of Desalinated Seawater (m^3)

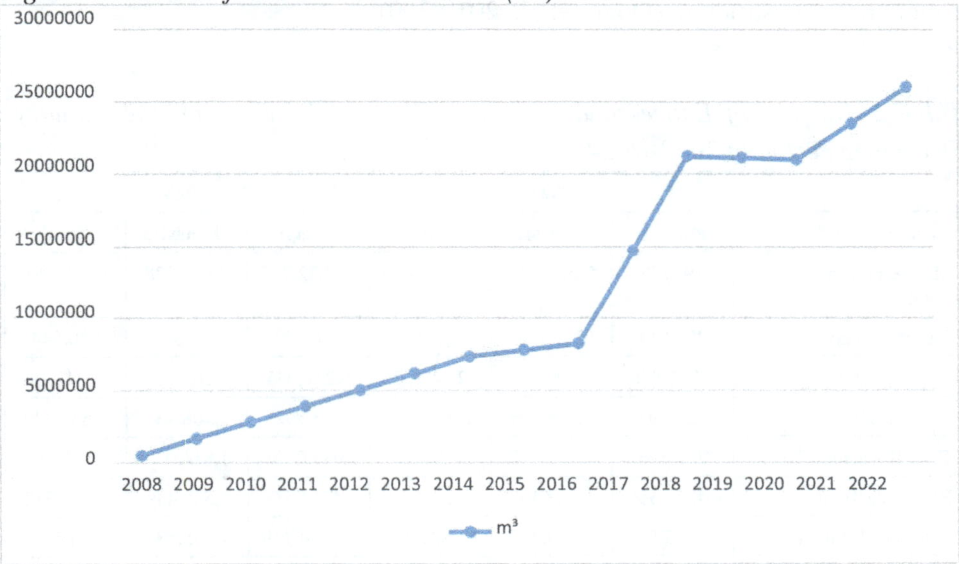

Source: FAO Aquastat Water Use Statistics, 2023.

Decreasing rainfall is a threat that could deepen the water crisis. To overcome this crisis, desalination could be used as a solution. According to calculations made in Türkiye, the use of desalinated seawater, which was around 500·000 ^{m3} at the beginning of the process·rose to 2.6 million ^{m3} by 2022. The growing population and production will increasingly drive up water demand. In order to meet water demand, desalination of seawater, a solution that Türkiye has not resorted to much today, will emerge as an important alternative solution.

CLIMATE CHANGE AND TOURIST NUMBERS

Climate characteristics are extremely important for tourism. The tourism sector develops in places with climate characteristics suitable for the format demanded by tourists. In addition, climate is not a constant phenomenon in nature, which has a dynamic structure. Over time, various changes occur due to both natural processes and human factors. Some of these effects occur rapidly, while others are relatively slower. Climate change is occurring due to global warming. The structural characteristics of regions are differing in terms of both precipitation and warming. The tourism sector has a seriously fragile structure against climate change. Although the tourism sector tries to resist these changes, its level of fragility is higher than its level of resistance. Countries' income levels also affect these vulnerabilities. Vulnerability increases in low-income countries, while high-income countries have a more resilient structure (Dogru et al., 2019:300).

Table 2. Number of Entries and Overnight Stays by Province for the January-December Period of 2023- 2024

	2023			2024		
Region	**Foreign**	**Domestic**	**Total**	**Foreign**	**Domestic**	**Total**
Mediterranean Total	20 917 246	7,537,623	28,454,869	23,672,725	7,974,278	31,647,003
Eastern Anatolia Total	226 813	712,154	938,967	218,679	761,865	980,544
Aegean Total	5 952 878	6,990,627	12,943,505	6,239,035	7,203,163	13,442,198
Southeast Total	226,803	2,058,037	2,284,840	228,892	2,408,437	2,637,329
Central Anatolia Total	787,496	1,770,148	2,557,644	989,270	1,814,038	2,803,308
Black Sea Total	438,835	1,383,564	1,822,399	475,063	1,549,449	2,024,512
Marmara Total	10,012,993	6,809,417	16,822,410	10,724,119	7,295,534	18,019,653
Türkiye Total	38,563,064	27,261,570	65,824,634	42,547,783	29,006,764	71,554,547

Source: TURKSTAT Tourism Statistics, 2025.

Total tourist arrivals increased from 2023 to 2024. The Mediterranean, Aegean, and Marmara regions lead this increase. The expected long-term temperature increases may result in these increases causing climate vulnerabilities. High temperature increases may also harm regional competitiveness (Diffenbaugh & Giorgi, 2012, pp. 813–814). Although there are increases in tourism in the short term, whether this effect will continue in the long term is a matter of debate.

Climate change-related temperature increases may also lead to an extension of the tourism season. In addition to the possibility of a seasonal extension in the Mediterranean and Aegean regions, the increase in temperature may also contribute to the development of summer tourism in the Black Sea region. The warming that occurs in the fall season and the extreme anomalies expected in the summer season may lead to an extension of the tourism season and the possibility of continuing tourism activities in the fall period. In order to prevent vulnerabilities, awareness of the consequences that climate change may have on the tourism sector must be raised, and both future investments and the transformation of existing capacity must be carried out accordingly.

CLIMATE CHANGE AND FISHING

Climate change directly affects the fishing industry by causing multidimensional environmental changes in marine and inland water ecosystems, such as rising temperatures, ocean acidification, declining oxygen levels, and habitat loss. Rising sea surface temperatures are shifting species distribution northward or to deeper waters, disrupting reproduction and migration cycles, and reducing stock productivity, particularly in commercially valuable pelagic species. Similarly, the increase in extreme weather events, changes in plankton dynamics, and the proliferation of invasive species are reducing the predictability of fishing activities and causing economic fluctuations in the sector. Therefore, the effects of climate change on fisheries must be addressed not only as an ecological issue but also as a socioeconomic problem.

Table 3. Aquatic Product Statistics (thousand tons), 2015-2024

	2015	2016	2017	2018	2019	2020	2021	2022	2023	2024
Aquatic Products (Total)	672.2	588.7	630.8	628.6	836.5	785.8	799.8	849.8	1010.3	933.2
Hunting	431.9	335.3	354.3	314.1	463.2	364.4	328.2	335.0	454.1	356.1
Seafood	397.7	301.5	322.2	284.0	431.6	331.3	295.0	301.7	420.5	322.5
Domestic Aquatic Products	34.2	33.9	32.1	30.1	31.6	33.1	33.1	33.3	33.5	33.6
Aquaculture	240.3	253.4	276.5	314.5	373.4	421.4	471.7	514.8	556.3	577.1

Source: Ministry of Agriculture and Forestry, Fisheries Statistics, 2024,

Statistics on marine fish and other marine product fishing are obtained from the Monthly Large- Scale Fishery Survey and the Seasonal Small-Scale Fishery Survey, conducted jointly by the Ministry of Agriculture and Forestry and the Turkish Statistical Institute. Statistics on inland water products and aquaculture production are compiled by the Ministry of Agriculture and Forestry.

According to the table, seafood fishing declined from 397.7 thousand tons in 2015 to 322.5 thousand tons in 2024; there was a dramatic decline of -23.3% in production between 2023 and 2024 alone. This sharp decline is consistent with climate change-related factors observed in the Mediterranean and Black Sea basins, such as increases in sea surface temperatures, shifts in species distribution, and the proliferation of alien/invasive species (such as balloonfish and lionfish).

Aquaculture data from 2015–2024 supports the increasing vulnerability of Türkiye's fishing sector to climate change. The systematic decline observed in marine fishing represents the direct reflection of climate-induced ecosystem pressures on economic output, while the rapid growth in the aquaculture sector demonstrates the industry's efforts to adapt to climate change through controlled production models. In the future, it will be imperative to integrate fisheries policies with the scientifically based management of climate risks. Türkiye's sustainable fisheries can only be secured through ecosystem-based management, the strengthening of stock monitoring systems, and the development of aquaculture within environmental limits.

EMPIRICAL LITERATURE

Reviewing the literature is of great importance in studies conducted on a particular issue in academic writing. This review is necessary not only to avoid repetition but also to identify gaps in the literature. The table below lists previous studies conducted in the field of marine economics.

Table 4. Previous Studies in Marine Economics

Author	Study	Period	Scope	Model	Results
Yasser et. al. (2024)	The blue economy effects on EUROMED tourism: forecasting approach	2000-2019	Euromed Countries	Panel Data ARDL	GDP, aquaculture, and trade openness positively correlate with the size of the blue economy in the EUROMED region. A 1% increase in aquaculture production increases tourist arrivals by 2.3%. GDP per capita contributes 3% to the increase in tourist numbers. However, price increases and rising CO2 emissions reduce tourist numbers, deteriorating the quality of the region's marine environment. Finally, trade openness will increase tourist numbers by reducing trade barriers for environmentally friendly goods and services.
Suh & Pomeroy (2020)	Projected Economic Impact of Climate Change on Marine Capture Fisheries in the Philippines	2013-2060	Philippines	Computable General Equilibrium (CGE)-Future Prediction Model-RCP 2.6-RCP 8.5	According to the projection results, extreme climate anomalies will have negative impacts on both fisheries and other economic variables. The fisheries sector's contribution to GDP is expected to decline by 9.27% in the optimistic scenario A and by 17.65% in the pessimistic scenario B by 2060.

continued on following page

Table 4. Continued

Author	Study	Period	Scope	Model	Results
Pazienza (2015)	The relationship between CO2 and Foreign Direct Investment in the agriculture and fishing sector of OECD countries: Evidence and policy considerations	1981-2005	30 OECD Countries	Panel Data Regression	A 1% increase in direct foreign investment in the agriculture and fisheries sector reduces carbon emissions by 0.0848%. According to this result, an increase in direct foreign investment leads to positive environmental outcomes.
Ahammed et. al. (2025)	Impact of blue economy factors on the sustainable economic growth of China	1980-2019	China	ARDL	In the study, a long-term cointegration between the variables was determined. TFF (Total Fishing Fleet), AFF (Aquaculture), AP (Agricultural Production), capital and trade variables have positive and significant effects on China's economic development.

continued on following page

Table 4. Continued

Author	Study	Period	Scope	Model	Results
Marwa et. al. (2024)	Determinants of the Blue Economy Growth in the Era of Sustainability: A Case Study of Indonesia	2012-2021	Indonesia	Panel Data Regression	The results show that information and communication technology (ICT) positively affects fisheries and aquaculture production and the share of the blue economy. Conversely, trade openness and electricity consumption have a negative impact on the share of the blue economy. The analysis also reveals that investment does not have a statistically significant effect on determining the share of the blue economy.
Martín-Cervantes et. al. (2025)	The potential of the Blue Economy to promote the generation of sustainable employment in the European Union	2009 - 2017	27 EU Member States	Panel Data Regression	According to the analysis, coastal tourism plays the most significant role in creating blue employment, followed by maritime transport and the exploitation of non-renewable marine resources. However, the fishing and aquaculture sectors exhibit an inverse relationship in job creation due to factors such as climate change and EU quota restrictions.

continued on following page

Table 4. Continued

Author	Study	Period	Scope	Model	Results
Su et. al. (2021)	Financial aspects of marine economic growth: From the perspective of coastal provinces and regions in China	2006 - 2016	Coastal Provinces and Regions in China	Bootstrap Panel Granger Causality	The relationship between financial development and maritime economic growth varies across regions due to differences in maritime industry structure, government policies, and financial market quality. In the eastern maritime economic environment, financial development appears to have a Granger causal effect on maritime economic growth. The developed financial system in this region allows maritime industries to fully utilize available financial resources, positively influencing economic growth.

continued on following page

Table 4. Continued

Author	Study	Period	Scope	Model	Results
Bhattacharya & Dash (2020)	Drivers of blue economy in Asia and Pacific Island countries: An empirical investigation of tourism and fisheries sectors	1996-2016	21 Asia and Pacific island countries	Panel Data Regression	According to the results of the study, a 1% increase in gross capital stock increases the size of the blue economy by 0.03%. A 1% depreciation of the local currency against the dollar also increases the number of tourists by 1.78%. In addition, as the level of trade openness increases, the size of the blue economy also grows.
Alharthi & Hanif (2020)	Impact of blue economy factors on economic growth in the SAARC countries	1995-2018	South Asian Association for Regional Cooperation (SAARC) Countries	Panel Data Regression	A 1% increase in the fishing industry in SAARC countries increases economic growth by 0.12%. The analysis results show that a 1% increase in aquaculture also stimulates economic growth by approximately 0.37%.

METHOD AND DATA SET

The study investigates the effect of marine stress variables (average sea surface temperature, sea water pH level), oil price, and exchange rate on the contribution of the blue economy to GDP. The ARDL cointegration test is used to determine the relationship between the variables. A series of tests must be performed to determine the structural characteristics of the data set.

Table 5. Variables Used in the Study and Their Descriptions

Variable	Abbreviation	Description	Unit	Source
Marine_GDP_Million_USD	LNMARINE	Blue economy contribution to GDP	Million USD	Turkish Statistical Institute + Central Bank of the Republic of Türkiye
Sea_Temperature_C	LNSST	Average sea surface temperature	°C	MGM/NOAA
Sea_pH	LNPH	Sea water acidity-alkalinity balance	pH	TÜBİTAK MAM
Oil_Price_USD	**LNOIL**	**Brent crude oil spot price**	**USD/Barrel**	**World Bank**
USD/TRY Exchange Rate	LNUSD	Annual average exchange rate	TRY	TCMB

Whether the series contain a unit root, and if so, the unit root levels, are very important for the model to be established. In order to apply cointegration tests, the series must contain a unit root.

Table 6. Unit Root Test Results

Series Name	Model		ADF	
			Statistics	Prob.
LNMARINE	Level	None	-0.733030	0.3868
		Constant	2.551450	0.9999
		Constant Trend	1.296808	0.9999
	First Difference	None	-0.256115	0.5817
		Constant	-0.687892	0.8291
		Constant Trend	-5.969017	0.0004***
LNSST	Level	None	1.054918	0.9170
		Constant	-5.622418	0.0002***
		Constant Trend	-1.652611	0.7359
	First Difference	None	-3.327691	0.0022***
		Constant	-1.400133	0.5602
		Constant Trend	-5.632326	0.0009***

continued on following page

Table 6. Continued

Series Name	Model		ADF	
			Statistics	Prob.
LNPH	Level	None	-8.687547	0.0000***
		Constant	1.606512	0.9989
		Constant Trend	-6.279232	0.0002***
	First Difference	None	-0.044386	0.6545
		Constant	-9.966568	0.0000***
		Constant Trend	-4.449877	0.0117**
LNOIL	Level	None	0.609749	0.8408
		Constant	-2.052666	0.2641
		Constant Trend	-1.879142	0.6325
	First Difference	None	-4.598707	0.0001***
		Constant	-4.653580	0.0014***
		Constant Trend	-4.718805	0.0056***
LNUSD	Level	None	4.247571	0.9999
		Constant	2.325996	0.9999
		Constant Trend	2.265835	1.0000
	First Difference	None	-0.328881	0.5546
		Constant	0.657372	0.9869
		Constant Trend	-5.366761	0.0014***

Note: *** indicates significance at the 1% level, ** indicates significance at the 5% level, and * indicates significance at the 1% level. The lag length criterion in ADF unit root tests is determined by the Schwarz information criterion.

According to the ADF results, the LN SST series is I(0) (stationary at the level), while the LN PH, LN OIL, and LN USD series are I(1) (stationary at the first difference). Due to their different levels of stationarity and the fact that most of the series are I(1), the use of the ARDL cointegration test would be appropriate for the model.

Table 7. Diagnostic Test Results for ARDL Models

Test	Probability & Statistics
R^2	0.9995
Breusch-Godfrey Serial Correlation LM Test	0.3598
Heteroskedasticity Test: Breusch-Pagan-Godfrey	0.9448
Histogram-Normality Test	0.5375
Ramsey RESET Test	0.9026

The table shows the standard Diagnostic Test Results for the ARDL model. These tests are crucial for understanding the extent to which the established econometric model meets its assumptions and, consequently, how reliable the model is. The results presented in Table 8 show that the established ARDL model meets all the basic econometric assumptions and is therefore statistically robust. All diagnostic test results indicate that the model is sound, as the corresponding probability values are above the 5% significance level. The model's results are reliable, and it is appropriate to interpret the long-term coefficients (such as the negative effect of SST) and draw policy conclusions based on these coefficients.

Figure 6. Cusum and Cusum Square Test Results

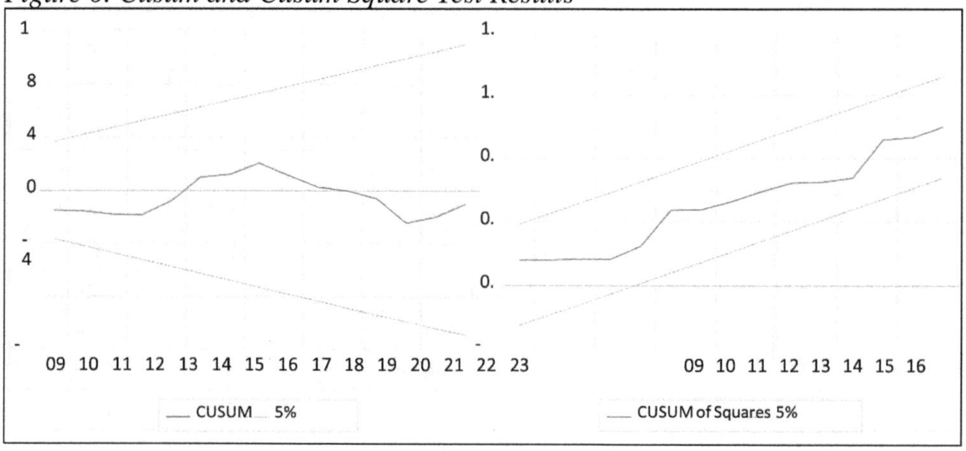

The fact that both the CUSUM and CUSUM of Squares graphs remain within the 5% significance limits strongly supports the conclusion that the coefficients of the established econometric model (ARDL) are stable, produce reliable estimates, and did not undergo any significant structural change during the estimation period.

These findings are fully consistent with the results obtained in the previous diagnostic tests indicating the absence of autocorrelation and heteroskedasticity issues, and reinforce the robustness of the model.

Table 8. F Bounds Test Results

Signif.	ARDL order	F-Statistics, Bounds Test
	4	6.897831
	I(0)	I(1)
10	2.2	3.09

continued on following page

Table 8. Continued

5	2.56	3.49
2.5	2.88	3.87
1	3.29	4.37

The calculated F-statistic value (6.897831) is greater than the Upper Bound I(1) values for all significance levels shown in the table: 6.897831 > 4.37 (1% level) - 6.897831 > 3.49 (5% level). This result indicates that there is a statistically significant long-term cointegration relationship between Marine Economy Value Added and Climate Stress Factors (SST, pH) and control variables.

The effects of climate change (SST and pH) are creating a permanent and long-term structural impact on Türkiye's marine economy, rather than short-term temporary shocks. This proves that interpreting long-term ARDL coefficients and developing policy recommendations is methodologically correct.

CONCLUSION

This study clearly demonstrates that climate change has multidimensional, profound, and lasting effects on Türkiye's blue economy. Climate projections indicate that the Mediterranean basin, where Türkiye is located, is warming faster than the global average and that temperature anomalies and precipitation declines will intensify in the coming decades. These physical changes directly affect marine ecosystems and human activities dependent on these ecosystems.

The results of econometric analysis prove that increases in sea surface temperature (SST) and ocean acidification (pH decline) have statistically significant and long-term negative effects on the blue economy GDP. This indicates that the impact of climate change on the blue economy is not a temporary shock but a structural problem.

In the tourism sector, although positive developments such as the extension of the tourism season due to warming have been observed in the short term, factors such as extreme heat waves, water scarcity, and forest fires in the long term pose a risk of reducing the competitiveness of traditional tourism regions such as the Mediterranean and Aegean. Tourism planning needs to be restructured to take these long- term climate risks into account.

In the fishing sector, the systematic decline observed in marine fishing from 2015 to 2024 reflects the devastating impact of climate change on marine biodiversity and stock productivity. In contrast, growth in the aquaculture sector points to the potential of controlled production models as an adaptation strategy to climate change.

Consequently, the sustainability of Türkiye's blue economy requires a holistic approach centered on climate change mitigation and adaptation strategies. In this context:

1. Ecosystem-Based Management: The restoration and protection of marine ecosystems and the sustainable management of fish stocks should be prioritized.
2. Scientific and Technological Infrastructure: Early warning systems and scientific research infrastructure should be strengthened to monitor the effects of climate change on the seas.
3. Sectoral Transformation: Climate-resilient, low-carbon investments in tourism should be encouraged; sustainable fishing practices should be promoted in fisheries, and aquaculture should be supported.
4. Policy Integration: Climate change should be addressed in an integrated manner with development, agriculture, energy, and water policies, and a policy framework encompassing all stakeholders of the blue economy should be established.

Only in this way can Türkiye reduce the systemic risks of climate change on the blue economy, protect marine resources, and build a sustainable economic future for future generations.

REFERENCES

Ahammed, S., Rana, M. M., Uddin, H., Majumder, S. C., & Shaha, S. (2025). Impact of blue economy factors on the sustainable economic growth of China. *Environment, Development and Sustainability*, *27*(3), 12625–12652. DOI: 10.1007/s10668-023-04411-6

Alharthi, M., & Hanif, I. (2020). Impact of blue economy factors on economic growth in the SAARC countries. *Maritime Business Review*, *5*(3), 253–269.

Alper, F. O., Golcek, A. G., & Alper, A. E. (2024). *The relationship between environmental taxes and renewable energy in Europe: A pathway to sustainable development?* (Vol. 2). Encyclopedia Of Monetary Policy, Financial Markets And Banking.

Bhattacharya, P., & Dash, A. K. (2020). *Drivers of blue economy in Asia and Pacific Island countries: an empirical investigation of tourism and fisheries sectors* (No. 1161). ADBI Working Paper Series.

Central Bank of the Republic of Türkiye. (2023). Exchange rates and currency statistics. https://www.tcmb.gov.tr/wps/wcm/connect/EN/TCMB+EN/Main+Menu/Statistics/Exchange+Rates

Chen, S., & Liu, Y. (2023). Exchange rate volatility and marine exports in developing countries: A panel NARDL approach. *The Journal of International Trade & Economic Development*, *32*(2), 245–267. DOI: 10.1080/09638199.2022.2052854

Demircan, M., Gürkan, H., Eskioğlu, O., Arabacı, H., & Coşkun, M. (2017). Climate change projections for Turkey: three models and two scenarios. *Turkish Journal of Water Science and Management*, *1*(1), 22-43.

Diffenbaugh, N. S., & Giorgi, F. (2012). Climate change hotspots in the CMIP5 global climate model ensemble. *Climatic Change*, *114*(3–4), 813–822. DOI: 10.1007/s10584-012-0570-x

Doyen, L., Bene, C., Bertignac, M., Blanchard, F., Cissé, A. A., Dichmont, C., & Péreau, J. C. (2022). Climate change and marine tourism in the Mediterranean: Economic impacts and adaptation strategies. *Tourism Management*, *89*, 104457. DOI: 10.1016/j.tourman.2021.104457

Fernandes, J. A., Cheung, W. W. L., Jennings, S., Butenschön, M., & Kay, S. (2021). Economic impacts of ocean acidification in Europe: A structural VAR approach. *Ecological Economics*, *179*, 106832. DOI: 10.1016/j.ecolecon.2020.106832

Food and Agriculture Organization of the United Nations. (2023). *AQUASTAT water use statistics* [Data set]. Retrieved September 15, 2025, from https://data.apps.fao.org/aquastat/?lang=en

France, W. W. F. (2021). *The climate change effect in the Mediterranean*. WWF France. [Access address: https://www.wwf.fr/sites/default/files/doc-2021-06/20210607_Rapport_The-Climate-Change-Effect-In-The-Mediterranean-Six-stories-from-an-overheating-sea_WWF-min.pdf]

General Directorate of Meteorology. (n.d.). *Climate change*. General Directorate of Meteorology. https://www.mgm.gov.tr/iklim/iklim-degisikligi.aspx

Global Ocean Acidification Observing Network. (2023). GOA-ON data portal: Mediterranean Sea pH measurements. http://portal.goa-on.org

IPCC. (2014). *Climate Change 2014: Synthesis Report. Contribution of Working Groups I, II and III to the Fifth Assessment Report of the Intergovernmental Panel on Climate Change* [R. K. Pachauri & L. A. Meyer (Eds.)]. IPCC.

Köktaş, A. M., Günel, T., & Gölçek, A. G. (2024). Environmental fiscal policy and carbon emissions in Turkey: Evidence from cointegration and causality analysis with structural breaks. *Tydskrif vir Studies in Ekonomie en Ekonometrie*, *48*(4), 365–375. DOI: 10.1080/03796205.2024.2389269

Lam, V. W. Y., Allison, E. H., Bell, J. D., Blythe, J., Cheung, W. W. L., Frölicher, T. L., Gasalla, M. A., & Sumaila, U. R.U. R. (2020). Climate change, tropical fisheries and prospects for sustainable development. *Nature Reviews. Earth & Environment*, *1*(9), 440–454. DOI: 10.1038/s43017-020-0071-9

Martín-Cervantes, P. A., Idowu, S. O., de Frutos Madrazo, P., & Frechoso Remiro, J. C. (2025). The potential of the Blue Economy to promote the generation of sustainable employment in the European Union. *Journal of Business and Socio-economic Development*, *5*(4), 355–371. DOI: 10.1108/JBSED-06-2025-0221

Martinez, M. L., Santos, C., & Fernandez, R. (2021). Non-linear effects in marine economy: Threshold regression models for critical temperature values. *Ecological Economics*, *188*, 107115. DOI: 10.1016/j.ecolecon.2021.107115

Merk, C., & Örnólfsson, S. (2019). Fuel price effects on fishing fleets: The case of Iceland, Norway, and Sweden. *Marine Policy*, *110*, 103543. DOI: 10.1016/j.marpol.2019.103543

Met Office Hadley Centre. (2018). *UKCP18 Guidance: Representative Concentration Pathways*. Retrieved from https://www.metoffice.gov.uk/binaries/content/assets/metofficegovuk/pdf/research/ukcp/ukcp18- guidance---representative-concentration-pathways.pdf

Ministry of Agriculture and Forestry. (2024). *Fisheries statistics, 2023-2024* [Data set]. https://cdniys.tarimorman.gov.tr/api/File/GetGaleriFile/482/DosyaGaleri/7951/tablo_1_su_urunleri_ist atistikleri_20232024.xlsx

National Oceanic and Atmospheric Administration. (2023). *Optimum Interpolation Sea Surface Temperature (OISST) dataset.* https://www.ncei.noaa.gov/products/optimum-interpolation-sst

O. (2022). Load capacity factor and financial globalization in Brazil: The role of renewable energy and urbanization. Frontiers in Environmental Science, 9, 689. DOI: 10.3389/fenvs.2021.823185

OECD. (2023). Maritime transport costs and fuel prices: A multilevel panel analysis of global sea trade. *OECD Transport Research Papers, 2023/2*. https://doi.org/DOI: 10.1787/5j8z6t9k8xwj-en

Pazienza, P. (2015). The relationship between CO2 and Foreign Direct Investment in the agriculture and fishing sector of OECD countries: Evidence and policy considerations. *Intelektinė ekonomika, 9*(1), 55-66.

Pesaran, M. H., Shin, Y., & Smith, R. J. (2001). Bounds testing approaches to the analysis of level relationships. *Journal of Applied Econometrics*, *16*(3), 289–328. DOI: 10.1002/jae.616

Republic of Türkiye Ministry of Environment. Urbanization and Climate Change, General Directorate of Meteorology. (2025). *Average temperature*. (Official statistics). [Access address: https://www.mgm.gov.tr/FILES/resmi-istatistikler/parametreAnalizi/2024-ortalama-sicaklik.pdf] Springer page (DOI verification): https://link.springer.com/article/10.1007/s10584-012-0570-x

Secretariat, U. F. M. (2019). *Risks associated with climate and environmental changes in the Mediterranean Region*. Union for the Mediterranean. [Access address: https://ufmsecretariat.org/wp-content/uploads/2019/10/MedECC-Booklet_EN_WEB.pdf]

Su, C. W., Song, Y., & Umar, M. (2021). Financial aspects of marine economic growth: From the perspective of coastal provinces and regions in China. *Ocean and Coastal Management*, *204*, 105550.

Suh, D., & Pomeroy, R. (2020). Projected economic impact of climate change on marine capture fisheries in the Philippines. *Frontiers in Marine Science*, 7, 232.

Sumaila, U. R., Tai, T. C., Lam, V. W. Y., Cheung, W. W. L., Bailey, M., & Zeller, D. (2021). Climate change impacts on marine fisheries: Economic impacts and adaptation strategies. *Marine Policy, 134*, 104795. DOI: 10.1016/j.marpol.2021.104795

TMMOB Chamber of Survey and Cadastre Engineers, 16, 3-6.

TÜBİTAK MAM. (2022). Economic effects of climate change in Turkish seas: Regional panel regression analysis. *TÜBİTAK Marmara Research Center Report, 2022/1*.

TURKSTAT. (2025). *Number of Entries and Overnight Stays by Province for the January-December Period of 2023–2024.* Turkish Statistical Institute. [Access address: https://data.tuik.gov.tr/Bulten/Index?p=Turizm-%C4%B0statistikleri-IV.-%C3%87eyrek:-Ekim- Aral%C4%B1k-ve-Y%C4%B1ll%C4%B1k,-2024-53660&dil=1]

Turkish Statistical Institute. (2023). Sectoral GDP accounts and marine economic activities. https://data.tuik.gov.tr

U.S. Energy Information Administration. (2023). Petroleum and other liquids: Brent crude oil prices. https://www.eia.gov/dnav/pet/pet_pri_spt_s1_d.htm

World Bank. (2023). Blue economy resilience to energy shocks: Evidence from small island developing states. World Bank Policy Research Working Paper, 10234. DOI: 10.1596/1813-9450-10234

World Bank. (2023). Commodity markets: Energy prices. https://www.worldbank.org/en/research/commodity-markets

World Bank. (2023). World Development Indicators: GDP by sector. https://databank.worldbank.org/source/world-development-indicators

Yavuz, E., Ergen, E., Avci, T., Akcay, F., & Kilic, E. (2024). Do the effects of aggregate and disaggregate energy consumption on different environmental quality indicators change in the transition to sustainable development? Evidence from wavelet coherence analysis. *Environmental Science and Pollution Research International, 31*(60), 67259–67279. DOI: 10.1007/s11356-023-30829-6

ENDNOTE

[1] The name given in European literature to the countries on the eastern shores of the Mediterranean starting in the 10th century.

Compilation of References

Abou Jaoude, J., & George Saade, R. (2019). Blockchain Applications – Usage in Different Domains. *IEEE Access : Practical Innovations, Open Solutions*, *7*, 45360–45381. DOI: 10.1109/ACCESS.2019.2902501

Abrams, M., & Weiss, J. (2008). *Malicious control system cyber security attack case study – Maroochy Water Services*. NIST Special Publication., https://www.mitre.org/sites/default/files/pdf/08_1145.pdf

Acciaro, M., Vanelslander, T., Sys, C., Ferrari, C., Roumboutsos, A., Giuliano, G., Lam, J. S. L., & Kapros, S. (2014). Environmental sustainability in seaports: A framework for successful innovation. *Maritime Policy & Management*, *41*(5), 480–500. DOI: 10.1080/03088839.2014.932926

Ahammed, S., Rana, M. M., Uddin, H., Majumder, S. C., & Shaha, S. (2025). Impact of blue economy factors on the sustainable economic growth of China. *Environment, Development and Sustainability*, *27*(3), 12625–12652. DOI: 10.1007/s10668-023-04411-6

Ajagunna, I., & Casanova, S. (2022). An analysis of the post-COVID-19 cruise industry: Could this be a new possibility for the luxury yacht sector in the Caribbean? *Worldwide Hospitality and Tourism Themes*, *14*(2), 115–123. DOI: 10.1108/WHATT-12-2021-0158

Aksu, H. H. (2021). Basin-based precipitation potential of Turkey. *Arabian Journal of Geosciences*, *14*(23), 2470. Advance online publication. DOI: 10.1007/s12517-021-08841-2

Al Mokdad, A. (2025). Inclusive and intelligent governance: Enabling sustainability through policy and technology. In *Social system reforms to achieve global sustainability* (pp. 173–238). IGI Global. DOI: 10.4018/979-8-3373-1280-4.ch008

Al Subhi, N. H., & Al Suqri, M. N. (2025, February). Understanding the role of artificial intelligence in marine spatial data infrastructure: A crucial step towards smarter ocean management. In *2025 International Conference for Artificial Intelligence, Applications, Innovation and Ethics (AI2E)* (pp. 1–6). IEEE. DOI: 10.1109/AI2E64943.2025.10983666

Alam, M., Khan, I. R., Siddiqui, F., & Alam, M. A. (2024). Artificial Intelligence as Key Enabler for Safeguarding the Marine Resources. In *Artificial Intelligence and Edge Computing for Sustainable Ocean Health* (pp. 409–451). Springer Nature Switzerland. DOI: 10.1007/978-3-031-64642-3_18

Albayrakoğlu, E. P. (2011). Climate change and security: The case for Turkey. *Gazi Akademik Bakış*, *5*(9), 59–75.

Alharthi, M., & Hanif, I. (2020). Impact of blue economy factors on economic growth in the SAARC countries. *Maritime Business Review*, *5*(3), 253–269.

Alhowaish, A. K. (2025). The blue economy in the Arabian Gulf: Trends, gaps, and pathways for sustainable coastal development. *Sustainability (Basel)*, *17*(19), 8809. Advance online publication. DOI: 10.3390/su17198809

Alper, F. O., Golcek, A. G., & Alper, A. E. (2024). *The relationship between environmental taxes and renewable energy in Europe: A pathway to sustainable development?* (Vol. 2). Encyclopedia Of Monetary Policy, Financial Markets And Banking.

Alves, B., Rigall-I-Torrent, R., Ballester, R., Benavente, J., & Ferreira, Ó. (2015). Coastal erosion perception and willingness to pay for beach management (Cadiz, Spain). *Journal of Coastal Conservation*, *19*(3), 269–280. DOI: 10.1007/s11852-015-0388-6

Amatya, B., & Khan, F. (2023). The efficacy of climatotherapy on medical rehabilitation outcomes: A systematic review. *The Journal of the International Society of Physical and Rehabilitation Medicine*, *6*(3), 47–59. DOI: 10.1097/ph9.0000000000000011

Amin, A. D. N., Rashdan, U., Mahadi, W. N. F. H. W., & Abdul, M. H. (2024). Environmental Injustice at Sea: The Impacts of Bilge Dumping on Indonesia's Coastal Communities. *International Journal of Research and Innovation in Social Science*, *8*(12), 3559–3568.

Antonelli, M., & Donelli, D. (2024). A narrative review of thalassotherapy and the health benefits of seawater and coastal climates. *International Journal of Biometeorology*, •••, 1–10.

Ap, J., & Crompton, J. L. (1998). Developing and testing a tourism impact scale. *Journal of Travel Research, 37*(2), 120–130. DOI: 10.1177/004728759803700203

Arabadzhyan, A., Figini, P., García, C., González, M. M., Lam-González, Y. E., & León, C. J. (2021). Climate change, coastal tourism, and impact chains–a literature review. *Current Issues in Tourism, 24*(16), 2233–2268. DOI: 10.1080/13683500.2020.1825351

Areche, F. O., Dominguez, J. A. J., Huaman, J. T., Yapias, R. J. M., Rivera, T. J. C., Perales, L. D. M., Quispe, J. D. D. H. T., Zea, C. Y. H., Galarza, C. R. C., Rodríguez, A. R., & Areche-Mansilla, D. Y. (2024). Maximizing the Potential of Marine Resources: A Sustainable Approach to High-Value Product Development from Seafood by-Products and Waste. *Current Research in Nutrition and Food Science, 12*(3), 1074–1092. DOI: 10.12944/CRNFSJ.12.3.7

Ashley, C., & Roe, D. (2001). Pro-poor tourism strategies: Making tourism work for the poor: A review of experience.

Avila, I. C., Kaschner, K., & Dormann, C. F. (2018). Current global risks to marine mammals: Taking stock of the threats. *Biological Conservation, 221*, 44–58. DOI: 10.1016/j.biocon.2018.02.021

Babu, C. S., Begum, M. A., & Karafolas, S. (2025). Blockchain and AI for advancing sustainability in the blue economy: Innovating marine resource management and environmental stewardship. In *FinTech for sustainable finance and a well-blue economy* (pp. 127–150). Springer Nature Switzerland. DOI: 10.1007/978-3-031-92390-6_7

Baihaqi, F., & Annida, S. B. (2024). The Social and Cultural Impact of Marine Tourism on Coastal Communities: A Qualitative Study at Tanjung Lesung Beach.

Bakker, K. (2022). Smart oceans: Artificial intelligence and marine protected area governance. *Earth System Governance, 13*, 100141. Advance online publication. DOI: 10.1016/j.esg.2022.100141

Bappenas. (2023). *Laporan Pembangunan Kelautan Indonesia*. Jakarta: Bappenas.

Barbosa-Filho, M. L. V., de Souza, G. B. G., de Faria Lopes, S., Hauser-Davis, R. A., Siciliano, S., & da Silva Mourão, J. (2020). Artisanal Fisher Knowledge and Attitudes Concerning Compressor Fishing in a North-Eastern Brazilian Marine Protected Area. *Human Ecology: an Interdisciplinary Journal, 48*(3), 357–366. DOI: 10.1007/s10745-020-00156-2

Bartolini, A., Di Gennaro, V., Reas, V., Mascolo, R. A., La Notte, A., Capriolo, A., & Ferrini, S. (2024). Ecosystem accounting for marine-based tourism provided by Posidonia oceanica in Italy. *One Ecosystem*, *9*, e129751. DOI: 10.3897/oneeco.9.e129751

Bashir, O. (2024). Harnessing Digitalization in Maritime Governance: A Systematic Review Strategic Challenges and Opportunities for Bangladesh. *Available at SSRN* 5125317.

Batool, S., Parveen, S., Javed, R., Saleem, M. S., Fatima, K., Ijaz, N., & Naseer, A. (2025). From overfishing to resilience: A global review of fisheries management strategies. *Haya: The Saudi Journal of Life Sciences*, *10*(8), 317–326.

Baum, T. (2007). Human resources in tourism: Still waiting for change. *Tourism Management*, *28*(6), 1383–1399. DOI: 10.1016/j.tourman.2007.04.005

Bayuningsih, D., Dzulkifli, M., & Muksin, M. (2024). Innovative Approaches in Tourism Development to Revitalize Global Sustainability. *International Journal Of Studies In Social Sciences And Humanities*, *1*(1), 1–17.

Bearzi, M. (2017). Impacts of marine mammal tourism. *Ecotourism's Promise and Peril: A Biological Evaluation*, 73-96.

Becker, D., Jungfer, C., & Track, T. (2019). Integrated industrial water management-challenges, solutions, and future priorities. *Chemieingenieurtechnik (Weinheim)*, *91*(10), 1367–1374. DOI: 10.1002/cite.201900086

Becker, G. S. (1993). *Human capital: A theoretical and empirical analysis, with special reference to education* (3rd ed.). University of Chicago Press., DOI: 10.7208/chicago/9780226041223.001.0001

Bello, A., Jahan, S., Farid, F., & Ahamed, F. (2023). A systemic review of the cybersecurity challenges in Australian water infrastructure management. *Water (Basel)*, *15*(1), 168. DOI: 10.3390/w15010168

Bethel, L., Jessen, H., & Hollander, J. (2021). Implementing the Port State Measures Agreement to combat illegal, unreported and unregulated fishing in the Caribbean. *Marine Policy*, *132*, 104643. DOI: 10.1016/j.marpol.2021.104643

Bhattacharya, P., & Dash, A. K. (2020). *Drivers of blue economy in Asia and Pacific Island countries: an empirical investigation of tourism and fisheries sectors* (No. 1161). ADBI Working Paper Series.

Bhola, R., Aggarwal, S., Dhawan, P., & Mečėjienė, G. (2025). Blue Economy and Big Data Analytics. In *FinTech for Sustainable Finance and a Well-Blue Economy* (pp. 197–209). Springer Nature Switzerland. DOI: 10.1007/978-3-031-92390-6_10

Biancalani, R., & Marinelli, M. (2021). Assessing SDG indicator 6.4.2 'level of water stress' at major basins level. *UCL Open Environment, 3*(5). Advance online publication. DOI: 10.14324/111.444/ucloe.000026 PMID: 37228788

Bianchi, R. V. (2003). Place and power in tourism development: Tracing the complex articulations of community and locality. *Pasos (El Sauzal), 1*(1), 13–32. DOI: 10.25145/j.pasos.2003.01.002

Bimonte, S., & Faralla, V. (2016). Does residents' perceived life satisfaction vary with tourist season? A two-step survey in a Mediterranean destination. *Tourism Management, 55*, 199–208. DOI: 10.1016/j.tourman.2016.02.011

Birdir, S., Ünal, Ö., Birdir, K., & Williams, A. T. (2013). Willingness to pay as an economic instrument for coastal tourism management: Cases from Mersin, Turkey. *Tourism Management, 36*, 279–283. DOI: 10.1016/j.tourman.2012.10.020

Bjelland, A. K. (2006). Selje Spa Thalasso-en helse-og velværekatedral?. *Tidsskrift for kulturforskning, 5*(2).

Boons, F., & Lüdeke-Freund, F. (2013). Business models for sustainable innovation: State of the art and steps towards a research agenda. *Journal of Cleaner Production, 45*, 9–19. DOI: 10.1016/j.jclepro.2012.07.007

Booth, H., Mourato, S., & Milner-Gulland, E. J. (2022). Investigating acceptance of marine tourism levies, to cover the opportunity costs of conservation for coastal communities. *Ecological Economics, 201*, 107578. DOI: 10.1016/j.ecolecon.2022.107578

Bottari, T., Mghili, B., Gunasekaran, K., & Mancuso, M. (2024). Impact of plastic pollution on marine biodiversity in Italy. *Water (Basel), 16*(4), 519. DOI: 10.3390/w16040519

Braidotti, L., Bertagna, S., Rappoccio, R., Utzeri, S., Bucci, V., & Marinò, A. (2023). On the inconsistency and revision of Carbon Intensity Indicator for cruise ships. *Transportation Research Part D, Transport and Environment, 118*, 103662. DOI: 10.1016/j.trd.2023.103662

Bramwell, B., & Lane, B. (2011). Critical research on the governance of tourism and sustainability. *Journal of Sustainable Tourism, 19*(4-5), 411–421. DOI: 10.1080/09669582.2011.580586

Brentan, B. M., Meirelles, G., Herrera, M., Luvizotto, E., & Izquierdo, J. (2017). Correlation Analysis of Water Demand and Predictive Variables for Short-Term Forecasting Models. Mathematical Problems in Engineering, 2017(1). DOI: 10.1155/2017/6343625

Brentan, B. M., Luvizotto, E.Jr, Herrera, M., Izquierdo, J., & Pérez-García, R. (2017). Hybrid regression model for near real-time urban water demand forecasting. *Journal of Computational and Applied Mathematics*, *309*, 532–541. DOI: 10.1016/j.cam.2016.02.009

Brida, J. G., & Zapata, S. (2010). Cruise tourism: Economic, socio-cultural and environmental impacts. *International Journal of Leisure and Tourism Marketing*, *1*(3), 205–226. DOI: 10.1504/IJLTM.2010.029585

Brown, S., Katz, D., Korotovskikh, D., & Kullman, S. (2024). Detecting Illegal, Unreported, and Unregulated Fishing through AIS Data and Machine Learning Approaches. *2024 Systems and Information Engineering Design Symposium (SIEDS)*, 319–324. DOI: 10.1109/SIEDS61124.2024.10534704

Bryhn, A., Kraufvelin, P., Bergström, U., Vretborn, M., & Bergström, L. (2020). A model for disentangling dependencies and impacts among human activities and marine ecosystem services. *Environmental Management*, *65*(5), 575–586. DOI: 10.1007/s00267-020-01260-1

Burgin, S., & Hardiman, N. (2015). Effects of non-consumptive wildlife-oriented tourism on marine species and prospects for their sustainable management. *Journal of Environmental Management*, *151*, 210–220. DOI: 10.1016/j.jenvman.2014.12.018

Busilacchi, S., Butler, J. R. A., Rochester, W. A., & Posu, J. (2018). Drivers of illegal livelihoods in remote transboundary regions: The case of the Trans-Fly region of Papua New Guinea. *Ecology and Society*, *23*(1), art46. DOI: 10.5751/ES-09817-230146

Butler, R. (2006). The future and the TALC. *The tourism area life cycle: Conceptual and theoretical issues, 2*, 281-290.

Buzzi, N. S., Menendez, M. C., Truchet, D. M., Delgado, A. L., & Severini, M. F. (2022). An overview on metal pollution on touristic sandy beaches: Is the COVID-19 pandemic an opportunity to improve coastal management? *Marine Pollution Bulletin*, *174*, 113275. DOI: 10.1016/j.marpolbul.2021.113275

Caporale, G. M., Claudio-Quiroga, G., & Gil-Alana, L. A. (2021). Analysing the relationship between CO2 emissions and GDP in China: A fractional integration and cointegration approach. *Journal of Innovation and Entrepreneurship*, *10*(1), 32. DOI: 10.1186/s13731-021-00173-5

Castilho, D., & Fuinhas, J. A. (2025). Exploring the effects of tourism capital investment on income inequality and poverty in the European Union countries. *Journal of Economic Structures*, *14*(1), 6. DOI: 10.1186/s40008-025-00349-2

Cavallini, I., Marzo, D., Scaccia, L., Scipioni, S., & Niccolini, F. (2023). Scuba diving tourism and the challenge of sustainability: Evidence from an explorative study in North African-Mediterranean countries. *EuroMed Journal of Business*, *20*(5), 1–26. DOI: 10.1108/EMJB-04-2022-0085

Cavole., ... (2025): Cavole, J. P., Cisneros-Montemayor, A. M., & Sumaila, U. R. (2025). Gender equality in the blue economy: Opportunities for inclusive governance in fisheries and aquaculture. In Sustainable blue economy transitions: Global perspectives (pp. 45–62). Springer.

CCICED. (2025): China Council for International Cooperation on Environment and Development. (2025). Sustainable blue economy under the vision of carbon neutrality [Special Policy Study Report]. https://cciced.eco/wp-content/uploads/2025/10/CCICED-2024-2025-Special-Policy-Study-Report-Sustainable-Blue-Economy-under-the-Vision-of-Carbon-Neutrality.pdf

Central Bank of the Republic of Türkiye. (2023). Exchange rates and currency statistics. https://www.tcmb.gov.tr/wps/wcm/connect/EN/TCMB+EN/Main+Menu/Statistics/Exchange+Rates

Charuka, B., Angnuureng, D. B., & Agblorti, S. K. (2023). Mapping and assessment of coastal infrastructure for adaptation to coastal erosion along the coast of Ghana. *Anthropocene Coasts*, *6*(1), 11. DOI: 10.1007/s44218-023-00026-6

Chen, G., Long, T., Bai, Y., & Zhang, J. (2019). A Forecasting Framework Based on Kalman Filter Integrated Multivariate Local Polynomial Regression: Application to Urban Water Demand. *Neural Processing Letters*, *50*(1), 497–513. DOI: 10.1007/s11063-019-10001-3

Chen, L., Chen, Z., Liu, Y., Lichfouse, E., Jiang, Y., Hua, J., Osman, A. I., Farghali, M., Huang, L., Zhang, Y., Rooney, D. W., & Yap, P. S. (2024). Benefits and limitations of recycled water systems in the building sector: A review. *Environmental Chemistry Letters*, *22*(2), 785–814. DOI: 10.1007/s10311-023-01683-2

Chenoweth, J., Hadjinicolaou, P., Bruggeman, A., Lelieveld, J., Levin, Z., Lange, M. A., Xoplaki, E., & Hadjikakou, M. (2011). Impact of climate change on the water resources of the eastern Mediterranean and Middle East region: Modeled 21st century changes and implications. *Water Resources Research, 47*(6), 2010WR010269. Advance online publication. DOI: 10.1029/2010WR010269

Chen, S., De Bruyne, C., & Bollempalli, M. (2020). Blue economy: Community case studies addressing the poverty–environment nexus in ocean and coastal management. *Sustainability (Basel), 12*(11), 4654. DOI: 10.3390/su12114654

Chen, S., & Liu, Y. (2023). Exchange rate volatility and marine exports in developing countries: A panel NARDL approach. *The Journal of International Trade & Economic Development, 32*(2), 245–267. DOI: 10.1080/09638199.2022.2052854

Chicco, D., Warrens, M. J., & Jurman, G. (2021). The coefficient of determination R-squared is more informative than SMAPE, MAE, MAPE, MSE and RMSE in regression analysis evaluation. *PeerJ. Computer Science, 7*, e623. DOI: 10.7717/peerj-cs.623 PMID: 34307865

Choi, E. C., Lee, J. S., & Chang, J. I. (2021). Willingness to pay for the prevention of beach erosion in Korea: The case of Haeundae beach. *Marine Policy, 132*, 104667. DOI: 10.1016/j.marpol.2021.104667

Chomka, M., & Gut-Winiarska, M. (2020). Prozdrowotne aspekty turystyki w środowisku morskim. Nauka i Społeczeństwo. *Przegląd nauk społecznych, technicznych, medycznych i pedagogicznych*, 1(1), 103-112.

Chuaysi, B., & Kiattisin, S. (2020a). Fishing Vessels Behavior Identification for Combating IUU Fishing: Enable Traceability at Sea. *Wireless Personal Communications, 115*(4), 2971–2993. DOI: 10.1007/s11277-020-07200-w

CISA. (2021a). Compromise of U.S. water treatment facility (Alert Code AA21-042A). Cybersecurity and Infrastructure Security Agency. https://www.cisa.gov/news-events/cybersecurity-advisories/aa21-042a

CISA. (2021b). Compromise of U.S. Water Treatment Facility. Retrieved from https://www.cisa.gov/news-events/cybersecurity-advisories/aa21-042a

CISA. (2025). Water and Wastewater Systems Sector. Retrieved from https://www.cisa.gov/topics/critical-infrastructure-security-and-resilience/critical-infrastructure-sectors/water-and-wastewater-sector

Cisneros-Montemayor, A. M., Becerril-García, E. E., Berdeja-Zavala, O., & Ayala-Bocos, A. (2020). Shark ecotourism in Mexico: Scientific research, conservation, and contribution to a Blue Economy. []. Academic Press.]. *Advances in Marine Biology*, *85*(1), 71–92.

CLIA (Cruise Lines International Association). (2020). *2020 State of the Cruise Industry Outlook*. https://cruising.org

Çolak, M. A., Öztaş, B., Özgencil, İ. K., Soyluer, M., Korkmaz, M., Ramírez-García, A., Metin, M., Yılmaz, G., Ertuğrul, S., Tavşanoğlu, Ü. N., Amorim, C. A., Özen, C., Apaydın Yağcı, M., Yağcı, A., Pacheco, J. P., Özkan, K., Beklioğlu, M., Jeppesen, E., & Akyürek, Z. (2022). Increased Water Abstraction and Climate Change Have Substantial Effect on Morphometry, Salinity, and Biotic Communities in Lakes: Examples from the Semi-Arid Burdur Basin (Turkey). *Water (Basel)*, *14*(8), 1241. DOI: 10.3390/w14081241

Cole, M. J., Bailey, R. M., Cullis, J. D. S., & New, M. G. (2018). Water for sustainable development in the Berg Water Management Area, South Africa. *South African Journal of Science*, *114*(3/4), 10. Advance online publication. DOI: 10.17159/sajs.2018/20170134

Coles, T. (2020). Impacts of climate change on tourism and marine recreation. *MCCIP Science Review*, •••, 593–615. DOI: 10.14465/2020.arc25.tre

Collins, M. (2008). Aspects of illegal, unreported and unregulated fishing the Southern Ocean - Rachel J. Baird Springer, Berlin, 2006 ISBN 978-1-4020-5338-2, 309 pages, £65.50. *Antarctic Science*, *20*(3), 309–309. DOI: 10.1017/S0954102008001223

Cooke, S. J., Twardek, W. M., Lennox, R. J., Zolderdo, A. J., Bower, S. D., Gutowsky, L. F., Danylchuk, A. J., Arlinghaus, R., & Beard, D. (2018). The nexus of fun and nutrition: Recreational fishing is also about food. *Fish and Fisheries*, *19*(2), 201–224. DOI: 10.1111/faf.12246

Corbett, J. J., & Winebrake, J. J. (2008). Emissions tradeoffs among alternative marine fuels: Total fuel cycle analysis of residual oil, marine gas oil, and marine diesel oil. *Journal of the Air & Waste Management Association*, *58*(4), 538–542. DOI: 10.3155/1047-3289.58.4.538

Culhane, F., Austen, M. C., Ashley, M., Javier, J., Kuit, S. H., Hung, N. P., Tran, H. D., Praptiwi, R. A., Sainal, S., Justine, E., Wulandari, P., Broszeit, S., Jontila, J. B., Johari, S., Creencia, L., Then, A. Y.-H., Gajardo, L. J., Maharja, C., Goh, H. C., & Langmead, O. (2024). Assessing impact risk to tropical marine ecosystems from human activities with a Southeast Asian example. *Journal of Applied Ecology*, *61*(12), 2897–2911. DOI: 10.1111/1365-2664.14812

Cunha, . (2025). Cunha, D. R., Ferreira, A. C., & Santos, J. P. (2025). Circular economy strategies for sustainable port management: Integrating reuse and digital optimization. *Maritime Policy & Management, 52*(3), 245–262. DOI: 10.1080/03088839.2024.2314567

Cunha-e-Sá, M. A., Dietrich, T., Faria, A., Nunes, L. C., Ortigão, M., Rosa, R., & Vieira-da-Silva, C. (2025). Willingness to pay for nature protection: Crowdfunding as a payment mechanism. *Environmental and Resource Economics, 88*(2), 367–396. DOI: 10.1007/s10640-024-00933-3

Cusack, C., Sethi, S. A., Rice, A. N., Warren, J. D., Fujita, R., Ingles, J., Flores, J., Garchitorena, E., & Mesa, S. V. (2021). Marine ecotourism for small pelagics as a source of alternative income generating activities to fisheries in a tropical community. *Biological Conservation, 261*, 109242. DOI: 10.1016/j.biocon.2021.109242

Dalkilic, H. Y., Bayçinar, M., & Samui, P. (2020). Evaluation of combined use of drought indices in the case of Konya closed basin. *International Journal of Global Warming, 23*(2), 169–190. DOI: 10.1504/IJGW.2021.112895

Dal, N., & Baysan, S. (2007). Coastal Land Use and Attitudes of Local People on Spatial Impacts of Tourism in Kuşadası. *Aegean Geography Journal, 16*(1-2), 69–85.

Danlu, L. (2023). A Study on the Talents Cultivation Model of Vocational Tourism English Students Based on the Integration of Industry and Education from the perspective of Stakeholders. *Frontiers in Educational Research, 6*(16).

Davis, D., & Tisdell, C. (1996). Economic management of recreational scuba diving and the environment. *Journal of Environmental Management, 48*(3), 229–248. DOI: 10.1006/jema.1996.0075

de Heer & Choudhury. (2025). de Heer, M., & Choudhury, S. (2025). Intelligent resilience in coastal adaptation: Integrating AI, IoT, and community-based monitoring for climate risk management. *Coastal Management, 53*(2), 89–107. DOI: 10.1080/08920753.2025.2345678

De la Cruz-Modino, R., & Cosentino, M. (2022). Conservation Hub: The Added Value of the Whale-Watching Industry. *Sustainability (Basel), 14*(20), 13471. DOI: 10.3390/su142013471

De Luna, X. (1998). Projected polynomial autoregression for prediction of stationary time series. *Journal of Applied Statistics, 25*(6), 763–775. DOI: 10.1080/02664769822756

De Melo Rodríguez, G., Martin-Alcalde, E., Murcia-González, J. C., & Saurí, S. (2017). Evaluating air emission inventories and indicators from cruise vessels at ports. *WMU Journal of Maritime Affairs*, *16*(3), 405–420. DOI: 10.1007/s13437-016-0122-8

Demircan, M., Gürkan, H., Eskioğlu, O., Arabacı, H., & Coşkun, M. (2017). Climate change projections for Turkey: three models and two scenarios. *Turkish Journal of Water Science and Management*, *1*(1), 22-43.

Diffenbaugh, N. S., & Giorgi, F. (2012). Climate change hotspots in the CMIP5 global climate model ensemble. *Climatic Change*, *114*(3–4), 813–822. DOI: 10.1007/s10584-012-0570-x

Dimmock, K., & Musa, G. (2015). Scuba Diving Tourism System: A framework for collaborative management and sustainability. *Marine Policy*, *54*, 52–58. DOI: 10.1016/j.marpol.2014.12.008

Dimopoulos, D., Queiros, D., & Zyl, C. V. (2025). Navigating the ecological decline of coral reef ecosystems and its impact on dive tourism in the East African Marine Ecoregion. *Tourism in Marine Environments*, *20*(2), 149–168. DOI: 10.3727/216901925X17460192677311

Do, Ó. A., & Seiz R. (2021). Effects of climate change in the Iberian Peninsula. Less rain and more uncertainty for the rivers. ANP/WWF and WWF Spain https://wwfes.awsassets.panda.org/downloads/efectos_cc_en_la_peninsula__2_.pdf?57520/Informe-sobre-los-efectos-del-cambio-climatico-en-la-Peninsula-Iberica

Dogan, H. Z. (1989). Forms of adjustment: Sociocultural impacts of tourism. *Annals of Tourism Research*, *16*(2), 216–236. DOI: 10.1016/0160-7383(89)90069-8

Doherty, P. D., Atsango, B. C., Ngassiki, G., Ngouembe, A., Bréheret, N., Chauvet, E., Godley, B. J., Machin, L., Moundzoho, B. D., Parnell, R. J., & Metcalfe, K. (2021). Threats of illegal, unregulated, and unreported fishing to biodiversity and food security in the Republic of the Congo. *Conservation Biology*, *35*(5), 1463–1472. DOI: 10.1111/cobi.13723 PMID: 33615559

Dong, Y. (2025). The impact of tourists on the marine environment: A review and managerial implications. *Regional Environmental Change*, *25*(2), 49. DOI: 10.1007/s10113-025-02385-x

Doxey, G. V. (1975). A causation theory of visitor–resident irritants: Methodology and research inferences. In *Proceedings of the Travel Research Association Sixth Annual Conference*, San Diego, CA, 195–198.

Doyen, L., Bene, C., Bertignac, M., Blanchard, F., Cissé, A. A., Dichmont, C., & Péreau, J. C. (2022). Climate change and marine tourism in the Mediterranean: Economic impacts and adaptation strategies. *Tourism Management*, *89*, 104457. DOI: 10.1016/j.tourman.2021.104457

Drammeh, O. (2000). *Illegal.* Unreported And Unregulated Fishing In Small-Scale Marine And Inland Capture Fisheries.

Dreizis, Y. (2020). Water quality management in the coastal zone of the sea. In *E3S Web of Conferences* (Vol. 215, p. 04003). EDP Sciences.

Duraimutharasan, N., Deepan, A., Swadhi, R., Velmurugan, P. R., & Varshney, K. R. (2025). Enhancing control engineering through human-machine collaboration: AI for improved efficiency and decision-making. In *Harnessing AI for control engineering* (pp. 155–176). IGI Global. DOI: 10.4018/979-8-3693-7812-0.ch008

Ekpemuaka, E., Odunlade, O. R., & Maiyaki, S. F. (2025). Resilience in Africa's transition to a sustainable digital and blue economy: Policy and regulatory framework in Nigeria. In *Securing sustainable futures through blue and green economies* (pp. 149–172). IGI Global. DOI: 10.4018/979-8-3693-7893-9.ch006

Elston, J., Pinto, H., & Nogueira, C. (2024). Tides of change for a sustainable blue economy: A systematic literature review of innovation in maritime activities. *Sustainability (Basel)*, *16*(24), 11141. Advance online publication. DOI: 10.3390/su162411141

Eluri, P. R. R., Annapurnaiah, K., Uday Bhaskar, T. V. S., Kiran Kumar, N., Padmanabham, J., Karthik, T. N. C., Vighneshwar, S. P., Reddem, V. S., Venugopala Rao, V., Murali Krishna, A., Ganesh, M. V., Lotliker, A., Nagaraja Kumar, M., Francis, P. A., Joseph, S., Balakrishnan Nair, T. M., & Srinivasa Kumar, T. (2025). Empowering the Blue Economy: The role of ICT in ocean observations, data, information and advisory services. *CSI Transactions on ICT*, *13*(1), 3–25. DOI: 10.1007/s40012-025-00407-x

Enriquez-Acevedo, T., Botero, C. M., Cantero-Rodelo, R., Pertuz, A., & Suarez, A. (2018). Willingness to pay for Beach Ecosystem Services: The case study of three Colombian beaches. *Ocean and Coastal Management*, *161*, 96–104. DOI: 10.1016/j.ocecoaman.2018.04.025

European Commission. (2023). Directive (EU) 2022/2555 on measures for a high common level of cybersecurity across the Union (NIS2 Directive). Official Journal of the European Union. https://eur-lex.europa.eu/eli/dir/2022/2555/oj

Expósito, A. (2019). Valuing Households' Willingness to Pay for Water Transfers from the Irrigation Sector: A Case Study of the City of Seville (Southern Spain). *Sustainability (Basel), 11*(24), 6982. DOI: 10.3390/su11246982

Expósito, A., Espinosa, M., & Villa-Damas, A. (2021). Valuing visitor willingness to pay for urban green space conservation: Case of Maria Luisa Park in Seville, Spain. *Journal of Urban Planning and Development, 147*(3), 05021020. DOI: 10.1061/(ASCE)UP.1943-5444.0000700

Eyo, V. O., Eriegha, O. J., Andem, A. B., Joseph, A., & Ovioma, G. O. (2025). Harnessing the Blue Economy for Innovative Fisheries Management in Nigeria: Sustainable Strategies and Innovations. In *Handbook of Sustainable Blue Economy* (pp. 1–25). Springer Nature Switzerland. DOI: 10.1007/978-3-031-32671-4_88-1

FAO. (2022). *The State of World Fisheries and Aquaculture 2022*. FAO.

Faulkner, B., & Tideswell, C. (1997). A framework for monitoring community impacts of tourism. *Journal of Sustainable Tourism, 5*(1), 3–28. DOI: 10.1080/09669589708667273

Fernandes, J. A., Cheung, W. W. L., Jennings, S., Butenschön, M., & Kay, S. (2021). Economic impacts of ocean acidification in Europe: A structural VAR approach. *Ecological Economics, 179*, 106832. DOI: 10.1016/j.ecolecon.2020.106832

Fernández, P. R., & Rivera, A. (2024). SCUBA diver behavior, diver carrying capacity and its implications for sustainable reef management. *Marine Policy, 165*, 106199. DOI: 10.1016/j.marpol.2024.106199

Ferrary, M. (2024, November 25). The government begins the maintenance of the beaches along the entire coast of Malaga. La Opinión de Málaga. https://www.laopiniondemalaga.es/malaga/2024/11/25/gobierno-trabajos-mantenimiento-playas-malaga-112025799.html

Food and Agriculture Organization of the United Nations. (2023). *AQUASTAT water use statistics* [Data set]. Retrieved September 15, 2025, from https://data.apps.fao.org/aquastat/?lang=en

Food and agriculture organization. (2001). International plan of action to prevent, deter and eliminate illegal, unreported and unregulated fishing (2001). *Journal of International Wildlife Law & Policy, 4*(2), 185–201. DOI: 10.1080/13880290109353986

Food and Agriculture Organization. (2022). The State of World Fisheries and Aquaculture 2022. Towards Blue Transformation. In *FAO*. F.A.O., DOI: 10.4060/cc0461en

France, W. W. F. (2021). *The climate change effect in the Mediterranean*. WWF France. [Access address: https://www.wwf.fr/sites/default/files/doc-2021-06/20210607_Rapport_The-Climate-Change-Effect-In-The-Mediterranean-Six-stories-from-an-overheating-sea_WWF-min.pdf]

Freeman, A. M., Herriges, J. A., & Kling, C. L. (2014). *The Measurement of Environmental and Resource Values: Theory and Methods* (3rd ed.). Routledge. DOI: 10.4324/9781315780917

Fudge, M., Ogier, E., & Alexander, K. A. (2023). Marine and coastal places: Well-being in a blue economy. *Environmental Science & Policy*, *144*, 64–73. DOI: 10.1016/j.envsci.2023.03.002

Fuengirola City Council. (2023, March 22). Fuengirola achieves the 'S for Sustainability' certificate, from the Institute of Tourism Quality of Spain, on all its beaches. Fuengirola. https://www.fuengirola.es/fuengirola-logra-el-certificado-s-de-sostenibilidad-del-instituto-de-calidad-turistica-de-espana-en-la-totalidad-de-sus-playas/

Fuengirola City Council. (2024). Sustainability Plan. Department of Beaches. https://www.fuengirola.es/wp-content/uploads/2024/07/PLAN-SOSTENIBILIDAD-OGP.pdf

GAO, Government Accountability Office. (2024). Critical Infrastructure Protection: EPA Urgently Needs a Strategy to Address Cybersecurity Risks to Water Systems. GAO-24-106744. Retrieved from https://www.gao.gov/products/gao-24-106744

Garcés-Ordóñez, O., Díaz, L. F. E., Cardoso, R. P., & Muniz, M. C. (2020). The impact of tourism on marine litter pollution on Santa Marta beaches, Colombian Caribbean. *Marine Pollution Bulletin*, *160*, 111558. DOI: 10.1016/j.marpolbul.2020.111558

Garrod, B., & Wilson, J. C. (2004). Nature on the edge? Marine ecotourism in peripheral coastal areas. *Journal of Sustainable Tourism*, *12*(2), 95–120. DOI: 10.1080/09669580408667227

Gaspar, P., Henriques, V., Fonseca, P., Los, H., Cloarec, M. G., Da Silva, A. J., Da Silva, R. C. A. F., Grosso, N., & Campos, A. (2021). Monitoring Pelagic Fishing Activity in the Northeast Atlantic. *2021 IEEE International Geoscience and Remote Sensing Symposium IGARSS*, 5712–5715. DOI: 10.1109/IGARSS47720.2021.9553509

Gedik, S., & Mugan-Ertugral, S. (2019). The effects of marine tourism on water pollution. *Fresenius Environmental Bulletin*, *28*(2), 863–866.

Gee, R. O. W. (2025). Greening the blue ocean: Leading systemic transformation with regenerative intelligence. *Earth Environmental Science Research & Review, 8*(1), 01-27.

Geissdoerfer, M., Savaget, P., Bocken, N. M. P., & Hultink, E. J. (2017). The circular economy – A new sustainability paradigm? *Journal of Cleaner Production*, *143*, 757–768. DOI: 10.1016/j.jclepro.2016.12.048

General Directorate of Meteorology. (n.d.). *Climate change*. General Directorate of Meteorology. https://www.mgm.gov.tr/iklim/iklim-degisikligi.aspx

George, J., Janaki, L., & Parameswaran Gomathy, J. (2016). Statistical Downscaling Using Local Polynomial Regression for Rainfall Predictions – A Case Study. *Water Resources Management*, *30*(1), 183–193. DOI: 10.1007/s11269-015-1154-0

Gesami, B. (2024). Artificial intelligence-driven fisheries value chain optimization in Kenya's blue economy (Preprint). *SSRN*. https://ssrn.com/abstract=4892638 DOI: 10.2139/ssrn.4892638

Gesami, B. K., & Nunoo, J. (2024). Artificial intelligence in marine ecosystem management: Addressing climate threats to Kenya's blue economy. *Frontiers in Marine Science*, *11*, 1404104. Advance online publication. DOI: 10.3389/fmars.2024.1404104

Ghosh, O., & Kumar, B. (2025). An Overview of Maritime Research, Its Significance, Challenges, and Opportunities: Navigating the Future. *Research Methods for Advancing the Maritime Industry*, 71-94.

GhulamRabbany, M., Afrin, S., Rahman, A., Islam, F., & Hoque, F. (2013). Environmental effects of tourism. American Journal of Environment. *Energy and Power Research*, *1*(7), 117–130.

Gilliland, T. E., Sanchirico, J. N., & Taylor, J. E. (2020). Market-driven bioeconomic general equilibrium impacts of tourism on resource-dependent local economies: A case from the western Philippines. *Journal of Environmental Management*, *271*, 110968. DOI: 10.1016/j.jenvman.2020.110968

Ginting, A., & Jayanti, J. (2024). *Influence of communication and attitudes of foreign tourists towards local communities. Vernacular: Linguistics*. Literature, Communication and Culture Journal., DOI: 10.35447/vernacular.v3i2.827

Gissi, E., Manea, E., Mazaris, A. D., Fraschetti, S., Almpanidou, V., Bevilacqua, S., Coll, M., Guarnieri, G., Lloret-Lloret, E., Pascual, M., Petza, D., Rilov, G., Schonwald, M., Stelzenmüller, V., & Katsanevakis, S. (2021). A review of the combined effects of climate change and other local human stressors on the marine environment. *The Science of the Total Environment, 755*, 142564. DOI: 10.1016/j.scitotenv.2020.142564

Giupponi, C., Gain, A. K., & Farinosi, F. (2021). Spatial assessment of water use efficiency (SDG indicator 6.4.1) for regional policy support. *Frontiers in Environmental Science, 6*, 141. Advance online publication. DOI: 10.3389/fenvs.2018.00141

Gladstone, W., Curley, B., & Shokri, M. R. (2013). Environmental impacts of tourism in the Gulf and the Red Sea. *Marine Pollution Bulletin, 72*(2), 375–388. DOI: 10.1016/j.marpolbul.2012.09.017

Global Ocean Acidification Observing Network. (2023). GOA-ON data portal: Mediterranean Sea pH measurements. http://portal.goa-on.org

Gobler, C. J. (2020). Climate change and harmful algal blooms: Insights and perspective. *Harmful Algae, 91*, 101731. DOI: 10.1016/j.hal.2019.101731

Goh, J., Adepu, S., Junejo, K. N., & Mathur, A. (2017). A Dataset to Support Research in the Design of Secure Water Treatment Systems. In Havarneanu, G., Setola, R., Nassopoulos, H., & Wolthusen, S. (Eds.), Lecture Notes in Computer Science: Vol. 10242. *Critical Information Infrastructures Security. CRITIS 2016*. Springer., DOI: 10.1007/978-3-319-71368-7_8

Gomes, C. S., Fernandes, J. V., Fernandes, F. V., & Silva, J. B. (2021). Salt mineral water and thalassotherapy. In *Minerals latu sensu and Human Health: Benefits, Toxicity and Pathologies* (pp. 631–656). Springer International Publishing. DOI: 10.1007/978-3-030-65706-2_16

Gooden, A., Clarke, T. M., Meyer, L., & Huveneers, C. (2024). Wildlife tourism has little energetic impact on the world's largest predatory shark. *Animal Behaviour, 207*, 247–265. DOI: 10.1016/j.anbehav.2023.10.004

Goodman, . (2023). Goodman, B. A., Schurer, S., & Kark, S. (2023). Climate-resilient maritime infrastructure: Assessing risks to coastal systems from sea-level rise and extreme events. *Climatic Change, 176*(4), 45. DOI: 10.1007/s10584-023-03512-8

Görmüş, K. S., Kutoğlu, Ş. H., Şeker, D. Z., Özölçer, İ. H., Oruç, M., & Aksoy, B. (2014). Temporal analysis of coastal erosion in Turkey: A case study Karasu coastal region. *Journal of Coastal Conservation, 18*(4), 399–414. DOI: 10.1007/s11852-014-0325-0

Gössling, S., Hall, C. M., & Scott, D. (2018). Coastal and ocean tourism. *Handbook on marine environment protection: Science, impacts and sustainable management*, 773-790.

Gössling, S. (2002). Global environmental consequences of tourism. *Global Environmental Change*, *12*(4), 283–302. DOI: 10.1016/S0959-3780(02)00044-4

Goyal, K., & Goyal, S. (2024). Predicting PM2.5 Air Quality Using Random Forest Regression Enhanced with Polynomial Features. 2024 IEEE 12th International Conference on Intelligent Systems (IS), 1–6. DOI: 10.1109/IS61756.2024.10705219

Greenwood, D. J. (1989). Culture by the pound: An anthropological perspective on tourism as cultural commodification. In Smith, V. (Ed.), *Hosts and Guests: The Anthropology of Tourism* (pp. 171–185). University of Pennsylvania Press.

Griggs, G., & Reguero, B. G. (2021). Coastal adaptation to climate change and sea-level rise. *Water (Basel)*, *13*(16), 2151. DOI: 10.3390/w13162151

Guo, S., Feng, X., & Zhou, B. (2024). Trends and perspectives of marine sports tourism: A bibliometric analysis and systematic review. *Water* (20734441), 16(19).

Hagan, S. (2014). Too Big to Tackle? the Persistent Problem of Pirate Fishing and the New Focus on Port State Measures. *Suffolk Transnational Law Review*, *37*(109).

Hall, C. M. (2001). Trends in ocean and coastal tourism: The end of the last frontier? *Ocean and Coastal Management*, *44*(9-10), 601–618. DOI: 10.1016/S0964-5691(01)00071-0

Handmer, . (2024). Handmer, J., Nalau, J., & Vanuatu, P. (2024). Gender and climate resilience in Pacific Island fisheries: Empowering women for sustainable conservation. *Ocean and Coastal Management*, *251*, 107156. Advance online publication. DOI: 10.1016/j.ocecoaman.2024.107156

Han, F., Liu, Y., Tian, H., Li, J., & Tian, Y. (2025). A comprehensive framework incorporating deep learning for analyzing fishing vessel activity using automatic identification system data. *ICES Journal of Marine Science*, *82*(2), fsae166. Advance online publication. DOI: 10.1093/icesjms/fsae166

Hanley, N., & Barbier, E. (2009). *Pricing Nature: Cost-Benefit Analysis and Environmental Policy*. Edward Elgar.

Hayati, Y., Adrianto, L., Krisanti, M., Pranowo, W. S., & Kurniawan, F. (2020). Magnitudes and tourist perception of marine debris on small tourism island: Assessment of Tidung Island, Jakarta, Indonesia. *Marine Pollution Bulletin*, *158*, 111393. DOI: 10.1016/j.marpolbul.2020.111393

Hinis, M. A., & Geyikli, M. S. (2023). Accuracy evaluation of standardized precipitation index (SPI) estimation under conventional assumption in Yeşilırmak, Kızılırmak, and Konya closed basins, Turkey. *Advances in Meteorology*, *2023*, 1–13. Advance online publication. DOI: 10.1155/2023/5142965

Hoegh-Guldberg, O., Pendleton, L., & Kaup, A. (2019). People and the changing nature of coral reefs. *Regional Studies in Marine Science*, *30*, 100699. DOI: 10.1016/j.rsma.2019.100699

Hoekstra, A. Y., Mekonnen, M. M., Chapagain, A. K., Mathews, R. E., & Richter, B. D. (2012). Global monthly water scarcity: Blue water footprints versus blue water availability. *PLoS One*, *7*(2), e32688. Advance online publication. DOI: 10.1371/journal.pone.0032688 PMID: 22393438

Hu, S., Gao, J., Zhong, D., Deng, L., Ou, C., & Xin, P. (2021). An Innovative Hourly Water Demand Forecasting Preprocessing Framework with Local Outlier Correction and Adaptive Decomposition Techniques. *Water (Basel)*, *13*(5), 582. Advance online publication. DOI: 10.3390/w13050582

Hynes, S., Cawley, M., Deely, J., & Norton, D. (2024). Alternative approaches to measuring the value of tourism in marine and coastal areas in ocean economy accounting. *Marine Policy*, *168*, 106299. DOI: 10.1016/j.marpol.2024.106299

Iglesias-Campo., ... (2021): Iglesias-Campo, M., Sanjuán, M. C., & Iglesias, G. (2021). Marine and maritime spatial planning guide: Integrating stakeholder participation and gender equity. European Maritime Safety Agency (EMSA). https://www.emsa.europa.eu/publications/2021-marine-spatial-planning-guide.html

Ioras, F. (2025). Sustainable Management of Mediterranean Superyacht Marinas: A Comparative Assessment of Environmental Practices and Policy Implications. *Sustainability (Basel)*, *17*(4), 1377. DOI: 10.3390/su17041377

Ioras, F., & Bandara, I. (2025). Artificial Intelligence and Sustainable Practices in Coastal Marinas: A Comparative Study of Monaco and Ibiza. *Sustainability (Basel)*, *17*(16), 7404. DOI: 10.3390/su17167404

IPCC. (2014). *Climate Change 2014: Synthesis Report. Contribution of Working Groups I, II and III to the Fifth Assessment Report of the Intergovernmental Panel on Climate Change* [R. K. Pachauri & L. A. Meyer (Eds.)]. IPCC.

IPCC. (2019). Resumen para responsables de políticas. https://www.ipcc.ch/site/assets/uploads/sites/2/2019/09/IPCC-Special-Report-1.5-SPM_es.pdf

Issifu, I., Alava, J. J., Lam, V. W. Y., & Sumaila, U. R. (2022). Impact of Ocean Warming, Overfishing, and Mercury on European Fisheries: A Risk Assessment and Policy Solution Framework. *Frontiers in Marine Science*, *8*(770805), 1–13. DOI: 10.3389/fmars.2021.770805

Iyiola, A. O., Ogwu, M. C., & Izah, S. C. (2025). Artificial intelligence in fisheries: Transformative potentials and challenges. In *Artificial intelligence in fisheries* (pp. 1–23). CRC Press. DOI: 10.1201/9781003506287-1

Jambeck, J. R., Geyer, R., Wilcox, C., Siegler, T. R., Perryman, M., Andrady, A., ... & Law, K. L. (2015). Plastic waste inputs from land into the ocean. science, 347(6223), 768-771.

Jennings, G. (2007). Water-based tourism, sport, leisure, and recreation experiences. In *Water-based tourism, sport, leisure, and recreation experiences* (pp. 1–20). Routledge. DOI: 10.4324/9780080468310-7

Ji, X., & Ding, X. (2024). Analysis on the relationship between coastal tourism and marine pollution: An empirical analysis of China's 11 coastal regions. *Frontiers in Marine Science*, *11*, 1471467. DOI: 10.3389/fmars.2024.1471467

Jonas, L. C. (2024). Emerging Opportunities in Blue Tourism. *Tourism in Marine Environments*, *19*(1), 59–67. DOI: 10.3727/216901924X17117021040174

Kabil, M., Ali, M. A., Marzouk, A., & Dávid, L. D. (2024). Gender perspectives in tourism studies: A comparative bibliometric analysis in the MENA region. *Tourism Planning & Development*, *21*(3), 276–298. DOI: 10.1080/21568316.2022.2050419

Kaczynski, W. (2011). The future of blue economy: Lessons for European Union. *Foundations of Management*, *3*(1), 21–32. DOI: 10.2478/v10238-012-0033-8

Karatas-Ozkan, M., Tunalioglu, R., Ibrahim, S., Ozeren, E., Grinevich, V., & Kimaro, J. (2024). Actioning sustainability through tourism entrepreneurship: Women entrepreneurs as change agents navigating through the field of stakeholders. *Central European Management Journal*, *32*(1), 31–56. DOI: 10.1108/CEMJ-12-2021-0159

Kasim, M., Oetama, D., Bahar, A., Alelo, M., Wardani, K. K., & Susandari, H. (2021). The role of local communities in developing potential areas for marine tourism. [). IOP Publishing.]. *IOP Conference Series. Earth and Environmental Science*, *800*(1), 012055. DOI: 10.1088/1755-1315/800/1/012055

Kerr, J. T., & Ostrovsky, M. (2003). From space to species: Ecological applications for remote sensing. *Trends in Ecology & Evolution*, *18*(6), 299–305. DOI: 10.1016/S0169-5347(03)00071-5

Kester, J. G. (2003). Cruise tourism. *Tourism Economics, 9*(3), 337–350. DOI: 10.1177/135481660300900307

Khaskheli, M. B., Zhao, Y., & Lai, Z. (2025). Sustainable Maritime Governance of Digital Technologies for Marine Economic Development and for Managing Challenges in Shipping Risk: Legal Policy and Marine Environmental Management. *Sustainability (Basel), 17*(21), 9526. DOI: 10.3390/su17219526

Kim, A., Scott, C. P., & Swartz, W. (2024). Local perspectives on marine ecotourism development in a water-insecure island region: The case of Bocas del Toro, Panama. *Frontiers in Marine Science, 11*, 1377053. DOI: 10.3389/fmars.2024.1377053

Kim, D. E., & Lim, S. S. (2024). Economic impacts of the European Union carding system on global fish trade. *Marine Policy, 165*, 106208. DOI: 10.1016/j.marpol.2024.106208

Kim, M., & Boo, C. S. (2015). An exploratory study on visit intention of destination in marine health tourism. *Susan Haeyang Gyoyuk Yeon-Gu, 27*(1), 230–242. DOI: 10.13000/JFMSE.2015.27.1.230

Kim, Y. H., Howlett, E., Hwang, J. H., & Park, Y. G. (2014). Use of ship traffic density from Automatic Identification System (AIS) and Vessel Monitoring Systems (VMS) in marine spatial planning: A case study in New England coast, USA. *KMI International Journal of Maritime Affairs and Fisheries, 6*(1), 39–52. DOI: 10.54007/ijmaf.2014.6.1.39

Kizielewicz, J., & Luković, T. (2013). The phenomenon of the marina development to support the European model of economic development. *TransNav. International Journal on Marine Navigation and Safety Od Sea Transportation, 7*(3), 461–466. DOI: 10.12716/1001.07.03.19

Köktaş, A. M., Günel, T., & Gölçek, A. G. (2024). Environmental fiscal policy and carbon emissions in Turkey: Evidence from cointegration and causality analysis with structural breaks. *Tydskrif vir Studies in Ekonomie en Ekonometrie, 48*(4), 365–375. DOI: 10.1080/03796205.2024.2389269

Kolesnikova, M. (2020). EU maritime economy and COVID-19. *Sovremennaya Yevropa, 2020*(4), 102–111. DOI: 10.15211/soveurope42020102111

Kontogianni, A., Damigos, D., Tourkolias, Ch., Vousdoukas, M., Velegrakis, A., Zanou, B., & Skourtos, M. (2014). Eliciting beach users' willingness to pay for protecting European beaches from beachrock processes. *Ocean and Coastal Management, 98*, 167–175. DOI: 10.1016/j.ocecoaman.2014.06.019

Kothalawala, H. S., & Arachchi, R. S. S. W. (2021). Potential of developing scuba diving tourism in Sri Lanka: Scuba diving operators perspective. *Journal of Tourism Economics and Applied Research*, *5*(1), 10–28.

Kovačić, M., & Silveira, L. (2020). Cruise tourism: Implications and impacts on the destinations of Croatia and Portugal. *Pomorstvo*, *34*(1), 40–47. DOI: 10.31217/p.34.1.5

Kriström, B., & Riera, P. (1997). Contingent valuation method. Applications to the Spanish rural environment. *Economía Agraria*, *179*, 133–166.

Kumar, P., Jose, A. A., & Rajagopal, A. R. (2024). Sustainable Use of Marine Resources Using Ocean Technologies. In *Sustainable Development Goals* (pp. 249–274). CRC Press. DOI: 10.1201/9781003468257-14

Kurniawan, F., Adrianto, L., Bengen, D. G., & Prasetyo, L. B. (2023). Hypothetical effects assessment of tourism on coastal water quality in the Marine Tourism Park of the Gili Matra Islands, Indonesia. *Environment, Development and Sustainability*, *25*(8), 7959–7985. DOI: 10.1007/s10668-022-02382-8

Laghrissi, C. (2024, April 18). A NASA map warns of the "worrying" rise in sea level in Malaga. La Opinión de Málaga. https://www.laopiniondemalaga.es/malaga/2024/04/18/mapa-nasa-alerta-preocupante-crecida-nivel-mar-malaga-101234915.html

Lam, V. W. Y., Allison, E. H., Bell, J. D., Blythe, J., Cheung, W. W. L., Frölicher, T. L., Gasalla, M. A., & Sumaila, U. R.U. R. (2020). Climate change, tropical fisheries and prospects for sustainable development. *Nature Reviews. Earth & Environment*, *1*(9), 440–454. DOI: 10.1038/s43017-020-0071-9

Lan, N. T. P., & Thanh, D. T. N. (2023). Maritime security or economic growth: The choice of stakeholders in coastal tourism development in Binh Dai district, Ben Tre province, Vietnam. [). IOP Publishing.]. *IOP Conference Series. Earth and Environmental Science*, *1247*(1), 012005. DOI: 10.1088/1755-1315/1247/1/012005

Łapko, A., Strulak-Wójcikiewicz, R., Landowski, M., & Wieczorek, R. (2018). Management of waste collection from yachts and tall ships from the perspective of sustainable water tourism. *Sustainability (Basel)*, *11*(1), 121. DOI: 10.3390/su11010121

Larkin, K. E., Marsan, A. A., Tonné, N., Van Isacker, N., Collart, T., Delaney, C., & Calewaert, J. B. (2022). Connecting marine data to society. In *Ocean science data* (pp. 283–317). Elsevier. DOI: 10.1016/B978-0-12-823427-3.00003-7

Le Gouvello & Simard. (2024). Le Gouvello, C., & Simard, C.-H. (2024). Integrating blue carbon ecosystems into port development: Mangroves, seagrasses, and salt marshes for erosion control and sustainability. *Marine Policy*, *162*, 105987.

Lee, S., Telesetsky, A., & Schofield, C. (2014). 4 Slipping the Net: Why Is It So Difficult to Crack Down on IUU Fishing? In *Freedom of Navigation and Globalization* (pp. 88–119). Brill | Nijhoff. DOI: 10.1163/9789004284081_007

Lestari, D. I., Putra, A. R., & Larasuci, A. Y. (2020). The main consequences of continued illegal, unreported, and unregulated (IUU) fishing within Indonesian waters for maritime security actors and coastal communities. *Research. Social Development*, *9*(1), e24911566. DOI: 10.33448/rsd-v9i1.1566

Lewis, T. G. (2006). *Critical infrastructure protection in homeland security: Defending a networked nation*. John Wiley & Sons. DOI: 10.1002/0471789542

Li, D., & Fu, G. (2024). Deep Learning Model-Based Demand Forecasting for Secondary Water Supply in Residential Communities: A Case Study of Shanghai City, China. IEEE Access. DOI: 10.1109/ACCESS.2023.3288817

Liang, X. C., & Ding, D. S. (2002). Trends of overseas studies of ocean and coastal tourism. *Ziran Ziyuan Xuebao*, *17*(6), 783–791.

Li, L., Wu, B., & Patwary, A. K. (2022). How marine tourism promote financial development in sustainable economy: New evidences from South Asia and implications to future tourism students. *Environmental Science and Pollution Research International*, *29*(1), 1155–1172. DOI: 10.1007/s11356-021-15709-1

Liu, J., Li, X., Yang, H., Han, G., Liu, J., Zheng, C., & Zheng, Y. (2019). The water-energy nexus of megacities extends beyond geographic boundaries: A case of Beijing. *Environmental Engineering Science*, *36*(7), 778–788. Advance online publication. DOI: 10.1089/ees.2018.0553 PMID: 31346304

Liu, J., Zhao, Y., & Jang, S. C. (2021). Environmental perceptions and willingness to pay for preservation: Evidence from beach destinations in China. *International Journal of Tourism Research*, *23*(5), 792–804. DOI: 10.1002/jtr.2442

Liu, W., & Cao, Z. (2018). Positive role of marine tourism on economic stimulus in coastal area. *Journal of Coastal Research*, (83), 217–220.

Longépé, N., Hajduch, G., Ardianto, R., de Joux, R., Nhunfat, B., Marzuki, M. I., Fablet, R., Hermawan, I., Germain, O., Subki, B. A., Farhan, R., Muttaqin, A. D., & Gaspar, P. (2018). Completing fishing monitoring with spaceborne Vessel Detection System (VDS) and Automatic Identification System (AIS) to assess illegal fishing in Indonesia. *Marine Pollution Bulletin*, *131*, 33–39. DOI: 10.1016/j.marpolbul.2017.10.016 PMID: 29106935

Longo, C. S., Buckley, L., Good, S. D., Gorham, T. M., Koerner, L., Lees, S., Liow, S. Y., Oloruntuyi, O., Schley, D., Rice, J., & Currey, R. J. C. (2021). A Perspective on the Role of Eco-Certification in Eliminating Illegal, Unreported and Unregulated Fishing. *Frontiers in Ecology and Evolution*, *9*, 637228. Advance online publication. DOI: 10.3389/fevo.2021.637228

Lopez-Rivas, J. D., & Cardenas, J. C. (2024). What is the economic value of coastal and marine ecosystem services? A systematic literature review. *Marine Policy*, *161*, 106033. DOI: 10.1016/j.marpol.2024.106033

Luiijf, H., & Paske, B. (2015). Cyber Security of Industrial Control Systems..

MacCannell, D. (1976). *The Tourist: A New Theory of the Leisure Class*. University of California Press.

Machado, J. T. M., & de Andrés, M. (2023). Implications of offshore wind energy developments in coastal and maritime tourism and recreation areas: An analytical overview. *Environmental Impact Assessment Review*, *99*, 106999. DOI: 10.1016/j.eiar.2022.106999

Maione, C. (2021). Quantifying plastics waste accumulations on coastal tourism sites in Zanzibar, Tanzania. *Marine Pollution Bulletin*, *168*, 112418. DOI: 10.1016/j.marpolbul.2021.112418

Marinelli, M., Biancalani, R., Joyce, B., & Djihounessi, M. B. (2025). A New Methodology to Estimate the Level of Water Stress (SDG 6.4.2) by Season and by Sub-Basin Avoiding the Double Counting of Water Resources. *Water (Basel)*, *17*(10), 1543. Advance online publication. DOI: 10.3390/w17101543

Martín-Cervantes, P. A., Idowu, S. O., de Frutos Madrazo, P., & Frechoso Remiro, J. C. (2025). The potential of the Blue Economy to promote the generation of sustainable employment in the European Union. *Journal of Business and Socio-economic Development*, *5*(4), 355–371. DOI: 10.1108/JBSED-06-2025-0221

Martínez Vázquez, R. M., Milán García, J., & De Pablo Valenciano, J. (2021). Analysis and Trends of Global Research on Nautical, Maritime and Marine *Tourism. Journal of Marine Science and Engineering*, *9*(1), 93. DOI: 10.3390/jmse9010093

Martinez, M. L., Santos, C., & Fernandez, R. (2021). Non-linear effects in marine economy: Threshold regression models for critical temperature values. *Ecological Economics*, *188*, 107115. DOI: 10.1016/j.ecolecon.2021.107115

Martinez-Mireles, J. R., Rodriguez-Flores, J., Figueroa-Diaz, R. A., Garcia-Escorza, B. B., Garcia-Marquez, M. A., & Austria-Cornejo, A. (2025). Integrating AI in green and blue infrastructure for sustainable smart cities. In *Smart water technology for sustainable management in modern cities* (pp. 337–364). IGI Global. DOI: 10.4018/979-8-3693-8074-1.ch014

Masni, H., Hutabarat, Z. S., Rahmadani, C., Yulita, K., Putri, R. A., Nurwati, N., Pianti, U. O., Safitri, R., Ramadani, N., Deskiana, I., Alfarizi, R. P., Putri, R. M., & Yuliana, E. (2024). Strategy for Implementing the Blue Economy Concept for Community Empowerment in Achieving Sustainable Development Goals. *International Journal of Multidisciplinary Sciences and Arts*, *3*(3), 37–43. DOI: 10.47709/ijmdsa.v3i3.3946

Mbaiwa, J. E. (2005). The socio-cultural impacts of tourism development in the Okavango Delta, Botswana. *Journal of Tourism and Cultural Change*, *2*(3), 163–185. DOI: 10.1080/14766820508668662

Mbavarira, T. M., & Grimm, C. (2021). A systemic view on circular economy in the water industry: Learnings from a Belgian and Dutch case. *Sustainability (Basel)*, *13*(6), 3313. Advance online publication. DOI: 10.3390/su13063313

McKegg, S. T. (1999). *Marine tourism in New Zealand: environmental issues and options* [Unpublished doctoral dissertation] University of Otago, New Zealand

McKercher, B., & Du Cros, H. (2002). *Cultural tourism: The partnership between tourism and cultural heritage management*. Routledge.

Medhekar, A. (2025). Gender-disaggregated data in marine governance: Challenges and strategies for policy evaluation in the blue economy. In *Gender and sustainability in coastal economies* (pp. 89–110). Routledge.

Meena, G., Vettriselvan, R., Rajesh, D., & Velmurugan, P. R. (2025). Diversity and Inclusion: Harnessing the Power of Inclusivity for Business Success. In *Security and Strategy Models for Key-Solving Institutional Frameworks* (pp. 203-234). IGI Global Scientific Publishing.

Mehrotra 2025Mehrotra, N. (2025). Economic contributions and challenges of the global blue economy: Livelihoods and ecosystem impacts. *Ocean and Coastal Management*, *245*, 107012. Advance online publication. DOI: 10.1016/j.ocecoaman.2025.107012

Mejjad, N., Laissaoui, A., Fekri, A., & Hammoumi, O. E. (2023). Marine plastic pollution in Morocco: State of the knowledge on origin, occurrence, fate, and management. *Environmental Science and Pollution Research International*, *30*(49), 107371–107389. DOI: 10.1007/s11356-023-26973-8

Mekonnen, M. M., & Hoekstra, A. Y. (2016). Four billion people facing severe water scarcity. *Sustainability*, *2*(2), e1500323. Advance online publication. DOI: 10.1126/sciadv.1500323 PMID: 26933676

Mendoza, A. A., Garret, T., Reyes, K. D. D., Soriano, P. A., & Cabauatan, R. (2021). The Impact of CO2 Emissions on the GDP per Capita, Employment Rate and Energy Consumption of China, Korea and Japan. [MJSSH]. *Malaysian Journal of Social Sciences and Humanities*, *6*(11), 315–333. DOI: 10.47405/mjssh.v6i11.1134

Merino, F., & Prats, M. A. (2022). Are blue flags a good indicator of the quality of sea water on beaches? An empirical analysis of the Western Mediterranean basin. *Journal of Cleaner Production*, *330*, 129865. DOI: 10.1016/j.jclepro.2021.129865

Merk, C., & Örnólfsson, S. (2019). Fuel price effects on fishing fleets: The case of Iceland, Norway, and Sweden. *Marine Policy*, *110*, 103543. DOI: 10.1016/j.marpol.2019.103543

Met Office Hadley Centre. (2018). *UKCP18 Guidance: Representative Concentration Pathways*. Retrieved from https://www.metoffice.gov.uk/binaries/content/assets/metofficegovuk/pdf/research/ukcp/ukcp18- guidance---representative-concentration-pathways.pdf

Middleton, V. T., & Hawkins, R. (1998). *Sustainable tourism: A marketing perspective*. Routledge.

Miller, M. L., & Auyong, J. (1991). Coastal zone tourism: A potent force affecting environment and society. *Marine Policy*, *15*(2), 75–99. DOI: 10.1016/0308-597X(91)90008-Y

Ministry of Agriculture and Forestry. (2024). *Fisheries statistics, 2023-2024* [Data set]. https://cdniys.tarimorman.gov.tr/api/File/GetGaleriFile/482/DosyaGaleri/7951/tablo_1_su_urunleri_ist atistikleri_20232024.xlsx

Mirziyoyeva, Z., & Salahodjaev, R. (2023). Renewable energy, GDP and CO2 emissions in high-globalized countries. *Frontiers in Energy Research*, *11*, 1123269. Advance online publication. DOI: 10.3389/fenrg.2023.1123269

Mitchell, R. C., & Carson, R. T. (1989). *Using Surveys to Value Public Goods: The Contingent Valuation Method*. Resources for the Future Press.

Mochklas, M., Maharani, R., Maretasari, R., Panggayudi, D. S., Oktaviani, M., & Muttaqin, R. (2024). Contribution Of Human Resources To Environmentally Friendly Entrepreneurial Models In Coastal Communities To Achieve Sustainable Development Goals (SDGs). *Journal of Lifestyle and SDGs Review*, *4*(2), 1–36. https://doi.org/https://doi.org/10.47172/2965-730X. DOI: 10.47172/2965-730X. SDGsReview.v4.n02.pe01770

Mochklas, M., Panggayudi, D. S., Hafidulloh, H., Iradawaty, S. N., & Pinaraswati, S. O. (2024). The Work Ethos of Kenjeran Beach Traders: A Study of Hard Work, Smart Work, and Sincere Work in the Coastal Tourism Industry. *Journal of Economics, Education, and Entrepreneurship*, *12*(1), 1–23. DOI: 10.26740/jepk.v12n1.p57-78

Mohanbabu, S., & Vettriselvan, R. (2025a). Focusing supply chain and container terminal challenges from the lens of Indian industries' perspectives with proposed data science interlinked artificial intelligence model. *International Journal of Procurement Management*, *24*(1), 92–114. DOI: 10.1504/IJPM.2025.148001

Mohanbabu, S., & Vettriselvan, R. (2025b). Will machine learning resolve the issues in container management in ports and logistics industries? *International Journal of Process Management and Benchmarking*, *20*(4), 559–575. DOI: 10.1504/IJPMB.2025.147399

Moric, I., Pekovic, S., Janinovic, J., Perovic, Đ., & Griesbeck, M. (2021). Cultural tourism and community engagement: Insight from Montenegro. Business Systems Research: *International journal of the Society for Advancing Innovation and Research in Economy*, 12(1), 164-178.

Musyafah, A., Prasetyo, M., Prananda, R., & Jati, O. (2025). Challenges in Implementing Sustainable Marine Tourism Policy against the Threat of Waste in Coastal Areas. *International Journal of Multidisciplinary Research And Analysis*. ..DOI: 10.47191/ijmra/v8-i01-20

Mutambara, P. D. E. (2025). Deploying AI to achieve UN SDGs: Details of the 17 goals. In *Deploying artificial intelligence to achieve the UN sustainable development goals: Enablers, drivers and strategic framework* (pp. 269–351). Springer Nature Switzerland. DOI: 10.1007/978-3-031-88423-8_12

Mycoo, M. A., & Gobin, J. F. (2013). Coastal management, climate change adaptation and sustainability in small coastal communities: Leatherback turtles and beach loss. *Sustainability Science*, *8*(3), 441–453. DOI: 10.1007/s11625-013-0212-x

Nababan, Y. N. (2023). Scuba diving as sustainability economic marine ecotourism: Scuba diving as sustainability economic marine ecotourism. *Journal of Oceanography and Aquatic Science*, *1*(2), 28–32. DOI: 10.56855/joane.v1i2.338

National Oceanic and Atmospheric Administration. (2023). *Optimum Interpolation Sea Surface Temperature (OISST) dataset.*_https://www.ncei.noaa.gov/products/optimum-interpolation-sst

Ndiaye, T. M. (2011). Illegal, Unreported and Unregulated Fishing: Responses in General and in West Africa. *Chinese Journal of International Law*, *10*(2), 373–405. DOI: 10.1093/chinesejil/jmr012

Nedelcu, A., Popescu, G., & Cioaca, A. (2015). Current diversity of cruise tourism. *Journal of Tourism Challenges and Trends*, *8*(1), 27–42.

Nesha Dushani, S., Aanesen, M., & Armstrong, C. W. (2023). Willingness to pay for mangrove restoration to reduce the climate change impacts on ecotourism in Rekawa coastal wetland, Sri Lanka. *Journal of Environmental Economics and Policy*, *12*(1), 19–32. DOI: 10.1080/21606544.2022.2065364

Nguyen, C., Schinckus, C., Su, T., & Chong, F. (2020). The Influence of Tourism on Income Inequality. *Journal of Travel Research*, *60*(7), 1426–1444. DOI: 10.1177/0047287520954538

Nguyen, D., Nguyen, M. K., Pham, M. T., Anh, N. T., Van Tri, D., Nguyen, H. L., & Nguyen, D. D. (2025). Investigating macro marine litter and beach cleanliness along Southern Vietnam beaches. *Marine Pollution Bulletin*, *212*, 117566. DOI: 10.1016/j.marpolbul.2025.117566

Nnanguma 2024Nnanguma, E. O. (2024). Institutionalizing gender inclusion in African marine governance: Pathways to legitimacy and ecosystem stability. *African Journal of Marine Science*, *46*(2), 145–162. DOI: 10.2989/1814232X.2024.2345678

Nyangon, J. (2025). Physics informed neural networks for maritime energy systems and blue economy innovations. *Machine Learning: Earth*, *1*(1), 011002.

O. (2022). Load capacity factor and financial globalization in Brazil: The role of renewable energy and urbanization. Frontiers in Environmental Science, 9, 689. DOI: 10.3389/fenvs.2021.823185

OECD. (2016). *Environmental impacts of cruise tourism*. OECD Publishing.

OECD. (2023). Maritime transport costs and fuel prices: A multilevel panel analysis of global sea trade. *OECD Transport Research Papers, 2023/2*. https://doi.org/DOI: 10.1787/5j8z6t9k8xwj-en

Olsson, G., & Lund, P. (2017). Water and Energy - Interconnections and Conflicts. 1(5), 1700056. https://doi.org/DOI: 10.1002/gch2.201700056

Onofri, L., & Nunes, P. A. (2013). Beach 'lovers' and 'greens': A worldwide empirical analysis of coastal tourism. *Ecological Economics, 88*, 49–56. DOI: 10.1016/j.ecolecon.2013.01.003

Onofrio, F., Rodella, I., & Gilli, M. (2025). Navigating WTP disparities: A study of tourist and resident perspectives on coastal management. *Frontiers in Environmental Economics, 4*, 1497532. DOI: 10.3389/frevc.2025.1497532

Orams, M. (2002). *Marine tourism: development, impacts and management.* Routledge. DOI: 10.4324/9780203197110

Otrachshenko, V., & Bosello, F. (2017). Fishing for answers? Impacts of marine ecosystem quality on coastal tourism demand. *Tourism Economics, 23*(5), 963–980. DOI: 10.1177/1354816616656422

Palma-Robles, M. A. (2009). *Combating IUU fishing: international legal developments.*

Papacharalampous, G., & Langousis, A. (2022). Probabilistic Water Demand Forecasting Using Quantile Regression Algorithms. *Water Resources Research, 58*(6), e2021WR030216. Advance online publication. DOI: 10.1029/2021WR030216

Papadaki, . (2025). Papadaki, L., Koundouri, P., & Seyhan, K. (2025). Bridging gaps for a thriving Black Sea blue economy: Insights from Greece and multi-sectoral forums. *Frontiers in Marine Science, 12*, 1491983. Advance online publication. DOI: 10.3389/fmars.2025.1491983

Papadaki, L., Halkos, G., & Koundouri, P. (2025). Circular investment principles in Greek port management: Socio-economic and environmental returns for sustainable blue growth. In *Sustainable maritime infrastructure: Investments and innovations* (pp. 112–135). Springer., DOI: 10.1007/978-3-031-56789-0_6

Papathanassis, A. (2020). The growth and development of the cruise sector: A perspective article. *Tourism Review, 75*(1), 130–135. DOI: 10.1108/TR-02-2019-0037

Papathanassis, A. (2023). A decade of 'blue tourism'sustainability research: Exploring the impact of cruise tourism on coastal areas. *Cambridge Prisms: Coastal Futures, 1*, e12. DOI: 10.1017/cft.2023.2

Pawson, M. G., Glenn, H., & Padda, G. (2008). The definition of marine recreational fishing in Europe. *Marine Policy, 32*(3), 339–350. DOI: 10.1016/j.marpol.2007.07.001

Pazienza, P. (2015). The relationship between CO_2 and Foreign Direct Investment in the agriculture and fishing sector of OECD countries: Evidence and policy considerations. *Intelektinė ekonomika, 9*(1), 55-66.

Pekpostalci, D. S., Tur, R., Mehr, A. D., Ghaffari, M. A. V., Dabrowska, D., & Nourani, V. (2023). Drought Monitoring and Forecasting across Turkey: A Contemporary Review. *Sustainability (Basel), 13*(7), 6080. Advance online publication. DOI: 10.3390/su15076080

Pérez-Juan, C. M., Prieto Fustes, L., Salvo Tierra, A. E., Navarro Jurado, E., Teixidor Costa, A., Danelutti, C., Iglesias Saugar, J., Wahlberg Macías, N., Jiménez Arce, J. A., & Velasco Martínez, L. C. (2022). *Tourism and the climate crisis on the Costa del Sol: Preliminary diagnosis and proposals for action. Gland, Switzerland: IUCN and Málaga.* University of Málaga.

Pesaran, M. H., Shin, Y., & Smith, R. J. (2001). Bounds testing approaches to the analysis of level relationships. *Journal of Applied Econometrics, 16*(3), 289–328. DOI: 10.1002/jae.616

Pessoa, R. M. C., Jiménez, J. A., da Costa, R. M., & Pereira, L. C. C. (2019). Federal conservation units in the Brazilian amazon coastal zone: An adequate approach to control recreational activities? *Ocean and Coastal Management, 178*, 104856. DOI: 10.1016/j.ocecoaman.2019.104856

Pham, T. (2020). Tourism in marine protected areas: Can it be considered as an alternative livelihood for local communities? *Marine Policy, 115*, 103891. DOI: 10.1016/j.marpol.2020.103891

Picken, F. (2025). Tourism and the blue economy. *Tourism Geographies, 27*(3-4), 538–546. DOI: 10.1080/14616688.2023.2291821

Pratiwi, R. A. I., Fani, L. A., & Kusasi, F. (2024). Blockchain Technology in Fisheries Industry: A Systematic Literature Review. *BIO Web of Conferences, 134*, 05004. DOI: 10.1051/bioconf/202413405004

Prihadi, D. J., Dhahiyat, A. P., Kristiadhi, F., & Pamungkas, W. (2024). Sustainable tourism in 2024: The Role of tourists and industry in Preserving marine nature. *MSJ: Majority Science Journal, 2*(4), 112–118. DOI: 10.61942/msj.v2i4.252

Radlińska, K., & Gardziejewska, B. (2022). The seasonal labor hoarding in tourist enterprises—Choice or necessity? *Sustainability (Basel), 14*(12), 6995. Advance online publication. DOI: 10.3390/su14126995

Raheja, S., Goyal, N., & Syan, A. S. (2025). Blockchain and AI: Pillars of sustainability in the blue economy. In *FinTech for sustainable finance and a well-blue economy* (pp. 151–169). Springer Nature Switzerland.

Ramdas, M., & Mohamed, B. (2014). Impacts of tourism on environmental attributes, environmental literacy and willingness to pay: A conceptual and theoretical review. *Procedia: Social and Behavioral Sciences*, *144*, 378–391. DOI: 10.1016/j.sbspro.2014.07.307

Rangel-Buitrago, N., Williams, A. T., Ergin, A., Anfuso, G., Micallef, A., & Pranzini, E. (2019). Coastal scenery: An introduction. *Coastal Scenery: Evaluation and Management*, 1-16.

Razak, M. A., Ghasemlouia, R., & Aksel, M. (2023). Tourism impact on coastal erosion: A case of Alanya. *International Journal of Environment and Geoinformatics*, *10*(4), 105–116. DOI: 10.30897/ijegeo.1287569

Rejón, R. (2024, June 6). Climate change eats up 45 metres of Spanish coastline weakened by construction in five years. ElDiario.es. https://www.eldiario.es/sociedad/cambio-climatico-come-cinco-anos-45-metros-costa-espanola-debilitada-construccion_1_11420555.html

Rengamani, J., James, F. A., Srinivasan, R., Poongavanam, S., & Vettriselvan, R. (2019). Impact on the usage of electronic data interchange (EDI) on the international shipping business in Chennai. *International Journal of Engineering and Advanced Technology*, *8*(2), 418–422.

Republic of Türkiye Ministry of Environment. Urbanization and Climate Change, General Directorate of Meteorology. (2025). *Average temperature*. (Official statistics). [Access address: https://www.mgm.gov.tr/FILES/resmi-istatistikler/parametreAnalizi/2024-ortalama-sicaklik.pdf] Springer page (DOI verification): https://link.springer.com/article/10.1007/s10584-012-0570-x

Rhoden, S., & Kaaristo, M. (2020). Liquidness: Conceptualising water within boating tourism. *Annals of Tourism Research*, *81*, 102854. DOI: 10.1016/j.annals.2019.102854

Roberts, C. M., O'Leary, B. C., McCauley, D. J., Cury, P. M., Duarte, C. M., Lubchenco, J., Pauly, D., Sáenz-Arroyo, A., Sumaila, U. R., Wilson, R. W., Worm, B., & Castilla, J. C. (2017). Marine reserves can mitigate and promote adaptation to climate change. *Proceedings of the National Academy of Sciences of the United States of America*, *114*(24), 6167–6175. DOI: 10.1073/pnas.1701262114

Rodella, I., Madau, F., Mazzanti, M., Corbau, C., Carboni, D., Utizi, K., & Simeoni, U. (2019). Willingness to pay for management and preservation of natural, semi-urban and urban beaches in Italy. *Ocean and Coastal Management*, *172*, 93–104. DOI: 10.1016/j.ocecoaman.2019.01.022

Santos, M. R., & Carvalho, L. C. (2025). AI-driven participatory environmental management: Innovations, applications, and future prospects. *Journal of Environmental Management*, *373*, 123864. Advance online publication. DOI: 10.1016/j.jenvman.2024.123864 PMID: 39752951

Saveriades, A. (2000). Establishing the social tourism carrying capacity for the tourist resorts of the east coast of the Republic of Cyprus. *Tourism Management*, *21*(2), 147–156. DOI: 10.1016/S0261-5177(99)00044-8

Schneider, F., Parsons, S., Clift, S., Stolte, A., & McManus, M. C. (2018). Collected marine litter—A growing waste challenge. *Marine Pollution Bulletin*, *128*, 162–174. DOI: 10.1016/j.marpolbul.2018.01.011

Schuhmann, P. W., Bangwayo-Skeete, P., Skeete, R., Seaman, A. N., & Barnes, D. C. (2024). Visitors' willingness to pay for ecosystem conservation in Grenada. *Journal of Sustainable Tourism*, *32*(8), 1644–1668. DOI: 10.1080/09669582.2023.2259119

Schuhmann, P., Skeete, R., Waite, R., Bangwayo-Skeete, P., Casey, J., Oxenford, H. A., & Gill, D. A. (2019). Coastal and marine quality and tourists' stated intention to return to Barbados. *Water (Basel)*, *11*(6), 1265. DOI: 10.3390/w11061265

Secretariat, U. F. M. (2019). *Risks associated with climate and environmental changes in the Mediterranean Region*. Union for the Mediterranean. [Access address: https://ufmsecretariat.org/wp-content/uploads/2019/10/MedECC-Booklet_EN_WEB.pdf]

Sen, A. (1999). *Capabilities and commodities*. Oxford University Press.

Sen, B., Topcu, S., Türkeş, M., & Warner, J. (2012). Projecting climate change, drought conditions and crop productivity in Turkey. *Climate Research*, *52*(1), 175–191. DOI: 10.3354/cr01074

Şentürk, H., Çil, C. Z., & Sağıroğlu, Ş. (2016). Siber Güvenlik Yatırım Kararları Üzerine Literatür İncelemesi. *Politeknik Dergisi*, *19*(1), 39–51.

Sevinç, F., & Güzel, T. (2017). Sustainable yacht tourism practices. *Management & Marketing Journal,* 15(1).

Sharpley, R. (2014). Host perceptions of tourism: A review of the research. *Tourism Management*, *42*, 37–49. DOI: 10.1016/j.tourman.2013.10.007

Shepperson, J. L., Hintzen, N. T., Szostek, C. L., Bell, E., Murray, L. G., & Kaiser, M. J. (2018). A comparison of VMS and AIS data: The effect of data coverage and vessel position recording frequency on estimates of fishing footprints. *ICES Journal of Marine Science*, *75*(3), 988–998. DOI: 10.1093/icesjms/fsx230

Simonsen, M., Gössling, S., & Walnum, H. J. (2019). Cruise ship emissions in Norwegian waters: A geographical analysis. *Journal of Transport Geography*, *78*, 87–97. DOI: 10.1016/j.jtrangeo.2019.05.014

Smith, M. K. (2015). *Issues in cultural tourism studies*. Routledge. DOI: 10.4324/9781315767697

Sodik, D. M. (2008). Non-legally binding international fisheries instruments and measures to combat illegal, unreported and unregulated fishing. *Australian International Law Journal, 15*(129).

Şorman, A. Ü., Mehr, A. D., & Hadi, S. J. (2018). Study on spatial-temporal variations of Meteorological-Agricultural droughts in Turkey. The International Archives of the Photogrammetry, Remote Sensing and Spatial Information Sciences, XLII(3/W4). DOI: 10.5194/isprs-archives-XLII-3-W4-483-2018

Spalding, M., Burke, L., Wood, S. A., Ashpole, J., Hutchison, J., & Zu Ermgassen, P. (2017). Mapping the global value and distribution of coral reef tourism. *Marine Policy*, *82*, 104–113. DOI: 10.1016/j.marpol.2017.05.014

Staffieri, E., de Lucia, G. A., Camedda, A., Poeta, G., & Battisti, C. (2019). Pressure and impact of anthropogenic litter on marine and estuarine reptiles: An updated "blacklist" highlighting gaps of evidence. *Environmental Science and Pollution Research International*, *26*(2), 1238–1249. DOI: 10.1007/s11356-018-3616-4

Stańczyk, J., Kajewska-Szkudlarek, J., Lipinski, P., & Rychlikowski, P. (2022). Improving short-term water demand forecasting using evolutionary algorithms. *Scientific Reports*, *12*(1), 13522. Advance online publication. DOI: 10.1038/s41598-022-17177-0 PMID: 35941276

Stepanets, I., & Hryniuk, D. (2021). *Economic and geographical aspects of the development of yachting as a promising sports tourism in the world*. Black Sea Economic Studies., DOI: 10.32843/bses.68-2

Stylidis, D. (2020). Exploring resident–Tourist interaction and its impact on tourists' destination image. *Journal of Travel Research*, *61*(1), 186–201. DOI: 10.1177/0047287520969861

Su, C. W., Song, Y., & Umar, M. (2021). Financial aspects of marine economic growth: From the perspective of coastal provinces and regions in China. *Ocean and Coastal Management, 204*, 105550.

Suh, D., & Pomeroy, R. (2020). Projected economic impact of climate change on marine capture fisheries in the Philippines. *Frontiers in Marine Science, 7*, 232.

Suherman, A., Santosa, M., Wijayanto, D., Sunoko, R., & Juwana, S. (2020). The Eradication of IUU Fishing in Indonesia for Economic Fisheries. *Saintek Perikanan : Indonesian Journal of Fisheries Science and Technology, 16*, 154–164.

Sukran, M. A., Kurniawan, T., Basri, H., & Furqan, A. (2025). Exploring marine tourism on sustainable development segment in Indonesia: SPAR-4-SLR. *Geo Journal of Tourism and Geosites, 58*(1), 243–256. DOI: 10.30892/gtg.58121-1406

Sultana, M., Bhattacharya, S., Adhikari, N., Sengupta, D., & De, D. (2024). Artificial intelligence for sustainable ocean health. In *Artificial intelligence and edge computing for sustainable ocean health* (pp. 3–38). Springer Nature Switzerland. DOI: 10.1007/978-3-031-64642-3_1

Sumaila, U. R., Tai, T. C., Lam, V. W. Y., Cheung, W. W. L., Bailey, M., & Zeller, D. (2021). Climate change impacts on marine fisheries: Economic impacts and adaptation strategies. *Marine Policy, 134*, 104795. DOI: 10.1016/j.marpol.2021.104795

Swadhi, R., Velmurugan, P. R., Gayathri, K., & Catherine, S. (2026). Evolving critical themes in advanced human resource management: Navigating change in the modern workplace. In *Critical aspects in advanced human resource management* (pp. 75–102). IGI Global.

Swarbrooke, J. (2020). *The Impact of Tourism on the Marine Environment*. Goodfellow Publishers Ltd.

Tandi, M. R., & Shrirao, N. M. (2023). AI-driven predictive analytics for sustainable stock assessment in coastal fisheries. *National Journal of Smart Fisheries and Aquaculture Innovation, 1*(1), 41–48.

Telfer, D. J., & Sharpley, R. (2015). *Tourism and development in the developing world*. Routledge. DOI: 10.4324/9781315686196

Tetreault, B. J. (2005). Use of the Automatic Identification System (AIS) for Maritime Domain Awareness (MDA). *Proceedings of OCEANS 2005 MTS/IEEE*, 1–5. DOI: 10.1109/OCEANS.2005.1639983

Thi, H. N., Thi, T. N., Trong, T. V., Duc, T. N., & Nghi, T. N. (2024). Sustainable tourism governance: A study of the impact of culture. *J. Gov. Regul, 13*(2).

Timothy, D. J. (2006). Empowerment and stakeholder participation in tourism destination communities. In *Tourism, power and space* (pp. 213–230). Routledge.

Timothy, D. J., & Boyd, S. W. (2003). *Heritage Tourism*. Pearson Education.

TMMOB Chamber of Survey and Cadastre Engineers, 16, 3-6.

Torres, R., & Momsen, J. H. (2004). Challenges and potential for linking tourism and agriculture to achieve pro-poor tourism objectives. *Progress in Development Studies*, *4*(4), 294–318. DOI: 10.1191/1464993404ps092oa

Tosun, C. (2000). Limits to community participation in the tourism development process in developing countries. *Tourism Management*, *21*(6), 613–633. DOI: 10.1016/S0261-5177(00)00009-1

Trice, A., Robbins, C., Philip, N., & Rumsey, M. (2021). *Challenges and opportunities for ocean data to advance conservation and management*. Ocean Conservancy. DOI: 10.15868/socialsector.43891

Tsai, F. M., Bui, T. D., Tseng, M. L., Lim, M. K., & Tan, R. R. (2021). Sustainable solid-waste management in coastal and marine tourism cities in Vietnam: A hierarchical-level approach. *Resources, Conservation and Recycling*, *168*, 105266. DOI: 10.1016/j.resconrec.2020.105266

TÜBİTAK MAM. (2022). Economic effects of climate change in Turkish seas: Regional panel regression analysis. *TÜBİTAK Marmara Research Center Report, 2022/1*.

Turkish Statistical Institute. (2023). Sectoral GDP accounts and marine economic activities. https://data.tuik.gov.tr

TURKSTAT. (2025). *Number of Entries and Overnight Stays by Province for the January-December Period of 2023–2024*. Turkish Statistical Institute. [Access address: https://data.tuik.gov.tr/Bulten/Index?p=Turizm-%C4%B0statistikleri-IV.-%C3%87eyrek:-Ekim-Aral%C4%B1k-ve-Y%C4%B1l1%C4%B1k,-2024-53660&dil=1]

TXOne Networks. (2025). Cyber Threats to Water and Wastewater Sector. Retrieved from https://www.txone.com/blog/cyber-threats-to-water-and-wastewater-sector/

U.S. Energy Information Administration. (2023). Petroleum and other liquids: Brent crude oil prices. https://www.eia.gov/dnav/pet/pet_pri_spt_s1_d.htm

UNEP. (2006). *Marine and coastal ecosystems and human wellbeing: A synthesis report based on the findings of the Millennium Ecosystem Assessment*. UNEP.

UNEP. (2009). *Sustainable Coastal Tourism: an integrated planning and management approach*. UNEP.

UNESCO. (2003). *Convention for the Safeguarding of the Intangible Cultural Heritage*. United Nations Educational, Scientific and Cultural Organization.

Unhasuta, S., Sasaki, N., & Kim, S. M. (2021). Impacts of tourism development on coastal communities in Cha-am Beach, the Gulf of Thailand, through analysis of local perceptions. *Sustainability (Basel)*, *13*(8), 4423. DOI: 10.3390/su13084423

United Nations Environment Programme. (2020). *Coral reefs*. Unep.Org. https://www.unep.org/topics/ocean-seas-and-coasts/blue-ecosystems/coral-reefs

Unnisa, Z., Govind, A., Lasserre, B., & Marchetti, M. (2023). Water Balance Trends along Climatic Variations in the Mediterranean Basin over the Past Decades. *Water (Basel)*, *15*(10), 1889. DOI: 10.3390/w15101889

UN-Water. (2022). *The United Nations world water development report 2022: Groundwater: Making the invisible visible*. UNESCO.

Van Phuong, T., & Pomeroy, R. S. (2023). Addressing Illegal, Unreported and Unregulated Fishing of Vietnamese Fishing Vessels in Foreign Waters. *Asian Fisheries Science*, *36*(1). Advance online publication. DOI: 10.33997/j.afs.2023.36.1.003

Van Putten, I., Metcalf, S., Frusher, S., Marshall, N., & Tull, M. (2014). Fishing for the impacts of climate change in the marine sector: A case study. *International Journal of Climate Change Strategies and Management*, *6*(4), 421–441. DOI: 10.1108/IJCCSM-01-2013-0002

Vanham, D., Alfieri, L., Flörke, M., Grimaldi, S., Lorini, V., de Roo, A., & Feyen, L. (2021). The number of people exposed to water stress in relation to how much water is reserved for the environment: A global modelling study. *The Lancet. Planetary Health*, *5*(11), e766–e774. DOI: 10.1016/S2542-5196(21)00234-5 PMID: 34774120

Vanham, D., Hoekstra, A. Y., Wada, Y., Bouraoui, F., de Roo, A., Mekonnen, M. M., van de Bund, W. J., Batelaan, O., Pavelic, P., Bastiaanssen, W. G. M., Kummu, M., Rockström, J., Liu, J., Bisselink, B., Ronco, P., Pistocchi, A., & Bidoglio, G. (2018). Physical water scarcity metrics for monitoring progress towards SDG target 6.4: An evaluation of indicator 6.4.2 "Level of water stress.". *The Science of the Total Environment*, *613–614*, 218–232. DOI: 10.1016/j.scitotenv.2017.09.056 PMID: 28915458

Velando, A., & Munilla, I. (2011). Disturbance to a foraging seabird by sea-based tourism: Implications for reserve management in marine protected areas. *Biological Conservation*, *144*(3), 1167–1174. DOI: 10.1016/j.biocon.2011.01.004

Venice, J. A., Vettriselvan, R., Jain, S., Madusudanan, K., & Aarthy, C. C. J. (2025). Performance Evaluation and Metrics in Blockchain Powered AI/ML: Data Analytics for Cognitive Internet of Things (CIoT). In Transforming Education With AI-Powered Personalized Learning (pp. 143-178). IGI Global Scientific Publishing.

Verma, S., & Sharma, M. (2024). Role of Digitalization in a Sustainable Blue Economy. International Journal of Teaching. *Learning and Education*, *4*(3), 618013.

Vetrimurugan, E., Shruti, V. C., Jonathan, M. P., Roy, P. D., Kunene, N. W., & Villegas, L. E. C. (2017). Metal concentration in the tourist beaches of South Durban: An industrial hub of South Africa. *Marine Pollution Bulletin*, *117*(1-2), 538–546. DOI: 10.1016/j.marpolbul.2017.02.036

Vettriselvan, . (2019). Vettriselvan, R., Suresh, A., & Kumar, P. (2019). Gender diversity and productivity in Indian maritime industries: Evidence from technical sectors. *Journal of Coastal Research*, *95*(sp1), 112–116.

Vitousek, S., Barnard, P. L., & Limber, P. (2017). Can beaches survive climate change? *Journal of Geophysical Research. Earth Surface*, *122*(4), 1060–1067. DOI: 10.1002/2017JF004308

Wada, Y., van Beek, L. P. H., & Bierkens, M. F. P. (2011). Modelling global water stress of the recent past: On the relative importance of trends in water demand and climate variability. *Hydrology and Earth System Sciences*, *15*(12), 3785–3808. DOI: 10.5194/hess-15-3785-2011

Wang, Y., Xiao, W., Wang, Y., Hou, B., Yang, H., Zhang, X., Yang, M., & Zhu, L. (2018). Exploring City Development Modes under the Dual Control of Water Resources and Energy-Related CO2 Emissions: The Case of Beijing, China. *Sustainability (Basel)*, *10*(9), 3155. Advance online publication. DOI: 10.3390/su10093155

Wang, Y., & Zhang, H. (2020). Influence of Cultural Identity on Tourists' Authenticity Perception and Destination Loyalty in Intangible Cultural Heritage Tourism. *Sustainability*, *12*(16), 6344. DOI: 10.3390/su12166344

Weatherdon, L. V., Magnan, A. K., Rogers, A. D., Sumaila, U. R., & Cheung, W. W. (2016). Observed and projected impacts of climate change on marine fisheries, aquaculture, coastal tourism, and human health: An update. *Frontiers in Marine Science*, *3*, 48. DOI: 10.3389/fmars.2016.00048

Welch, C., Paavilainen-Mäntymäki, E., Piekkari, R., & Plakoyiannaki, E. (2022). Reconciling theory and context: How the case study can set a new agenda for international business research. *Journal of International Business Studies*, *53*(1), 4–26. DOI: 10.1057/s41267-021-00484-5

West, M., & Harrison, J. (1989). Polynomial Trend Models (pp. 201–228). Yetik, A. K., Arslan, B., & Şen, B. (2023). Trends and Variability in Precipitation Across Turkey: A Multi-Method Statistical Analysis. https://doi.org/DOI: 10.1007/978-1-4757-9365-9_7

Williams 2023Williams, M. J. (2023). Gender equity in small-scale fisheries: Pathways to inclusion and resilience. *Fish and Fisheries*, *24*(5), 1123–1138. DOI: 10.1111/faf.12789

Williams, A. M., & Hall, C. M. (2000). Tourism and migration: New relationships between production and consumption. *Tourism Geographies*, *2*(1), 5–27. DOI: 10.1080/146166800363420

Wilson, J. C., & Garrod, B. (2003). Introduction. In Garrod, B., & Wilson, J. C. (Eds.), *Marine ecotourism: issues and experiences* (Vol. 7). Channel View Publications. DOI: 10.21832/9781873150436-004

World Bank. (2023). Blue economy resilience to energy shocks: Evidence from small island developing states. World Bank Policy Research Working Paper, 10234. DOI: 10.1596/1813-9450-10234

World Bank. (2023). Commodity markets: Energy prices. https://www.worldbank.org/en/research/commodity-markets

World Bank. (2023). World Development Indicators: GDP by sector. https://databank.worldbank.org/source/world-development-indicators

World Economic Forum. (2020). *Radically Reducing Plastic Pollution in Indonesia: Multi-Stakeholder Action Plan*. https://www.systemiq.earth/wp-content/uploads/2020/05/NPAP_Indonesia_Action_Roadmap_BahasaLow-1.pdf

World Tourism Organization (2024). *International Tourism Highlights, 2024 Edition*, November 2024, UN Tourism, Madrid, DOI:.DOI: 10.18111/9789284425808

Yang, J., Ryan, C., & Zhang, L. (2013). Social conflict in communities impacted by tourism. *Tourism Management*, *35*, 82–93. DOI: 10.1016/j.tourman.2012.06.002

Yavuz, E., Ergen, E., Avci, T., Akcay, F., & Kilic, E. (2024). Do the effects of aggregate and disaggregate energy consumption on different environmental quality indicators change in the transition to sustainable development? Evidence from wavelet coherence analysis. *Environmental Science and Pollution Research International*, *31*(60), 67259–67279. DOI: 10.1007/s11356-023-30829-6

Zaeri, N. (2025). Securing Sustainability Through Artificial Intelligence and Blue Economy. In *Securing Sustainable Futures Through Blue and Green Economies* (pp. 37–82). IGI Global Scientific Publishing. DOI: 10.4018/979-8-3693-7893-9.ch003

Zentner, Y., Margarit, N., Ortega, J., Casals, D., Medrano, A., Pagès-Escolà, M., & Linares, C. (2023). Marine protected areas in a changing ocean: Adaptive management can mitigate the synergistic effects of local and climate change impacts. *Biological Conservation*, *282*, 110048. DOI: 10.1016/j.biocon.2023.110048

Zetter, K. (2014). *Countdown to zero day: Stuxnet and the launch of the world's first digital weapon*. Crown.

Zhang, Z., Shan, Y., Zhao, D., Tillotson, M. R., Cai, B., Li, X., Zheng, H., Zhao, C., Guan, D., Liu, J., & Hao, Y. (2024). City level water withdrawal and scarcity accounts of China. *Scientific Data*, *11*(449), 449. Advance online publication. DOI: 10.1038/s41597-024-03115-4 PMID: 38702307

Zhao, D., Liu, J., Sun, L., Ye, B., Hubacek, K., Feng, K., & Varis, O. (2021). Quantifying economic-social-environmental trade-offs and synergies of water-supply constraints. *Water Research*, *195*, 116986. Advance online publication. DOI: 10.1016/j.watres.2021.116986 PMID: 33721677

Zheng, Y., & Liu, D. (2021). Research on marine pollution problems and solutions in China from the perspective of marine tourism. *Journal of Marine Science*, *3*(1). Advance online publication. DOI: 10.30564/jms.v3i1.2599

About the Contributors

Ali Gokhan Golcek Ph.D., is currently an Asst. Prof. in the Department of Public Finance at Niğde Ömer Halisdemir University, Türkiye. His scholarly interests span poverty, energy policy, public finance policy, health economics, and international political economy. He completed his PhD in Public Finance at Pamukkale University in 2022. Gölçek has published numerous articles and book chapters.

Bilal Göde is a Dr of Department of Public Finance at Pamukkale University, Turkey. His scholarly interests water economics, gerontological economics, public finance policy and international political economy. He completed his PhD in Public Finance at Pamukkale University in 2023. Göde has published numerous articles and book chapters.

***

Eşref Ay is an Independent Researcher, Dr., and he currently works as a tour guide. He completed his PhD thesis in tourism management, and his research interests are tourism guidance, sustainable tourism, tourism geography, and recreation. He has participated in some scientific projects related to sustainable tourism and tourism management.

Alfonso Expósito Professor (Associate) of the Department of Applied Economics (Economic Structure) at University of Malaga and member of the Water, Environmental and Agricultural Resources Economics (WEARE) research group. He has extensive research experience in microeconomic and macroeconomic analysis, with special emphasis in the field of water and environmental economics. He holds a Ph.D. in Economics (University of Seville, Spain) and a Master's degree in International Business (University College Dublin, Ireland).

Beyza Güdek is pursuing her academic career in the field of Management Information Systems (MIS) at the undergraduate, graduate, and doctoral levels. She received her bachelor's degree from Necmettin Erbakan University, her master's degree from Karadeniz Technical University, and is currently continuing her doctoral studies at Izmir Bakırçay University. As part of her Ph.D. program, she is also engaged in international academic training at Vilnius University. She currently serves as a research assistant in the Department of Management Information Systems at the Faculty of Economics and Administrative Sciences, Karadeniz Technical University. Her research interests include digital transformation, artificial intelligence, data analytics, and multi-criteria decision making.

Melik Güzel was working as a postdoctoral researcher on a project at Anadolu University. He is currently independent researcher. He has working experiences in tourism and health sector since 2008.

Sofiah Nur Iradawaty, S.E., M.M. was born in Surabaya on November 30, 1967. She has completed her studies at the Faculty of Economics in the IESP Department of Airlangga University Surabaya with a Bachelor of Economics (SE) degree in 1993. Then she continued her postgraduate studies in the Master of Management program at Wijaya Putra University Surabaya which she completed in 2008 with a Master of Management (MM) degree. She is a permanent lecturer in the Management study program at the Faculty of Economics, Yos Soedarso University Surabaya. She has been teaching in the Management study program at the Faculty of Economics, Yos Sudarso University Surabaya since 1995 until now. Some of the courses taught are Managerial Economics, Introduction to Macroeconomics, Introduction to Microeconomics, and Marketing Management. Previously, she taught at several educational institutions such as; PIKMI Surabaya, STIE YPM Sepanjang Sidoarjo, and at Mark Plus Professional Services (as an interviewer). He has also been a speaker and speaker in 2018 at the international seminar "Business Development Challenges & Strategies in the Industrial Revolution 4.0" and was a resource person in 2021 in the webinar "Sister Portfolio Implementation".

R. Srinivasan is Professor and Head, AMET Business School, AMET University, Chennai. With a Ph.D. in Management and multiple postgraduate qualifications, he brings over 32 years of academic and 6 years of industry experience. A seasoned administrator and scholar, he has guided numerous Ph.D., M.Phil., and MBA researchers and published widely in SCOPUS, UGC, and Web of Science journals. His research spans Finance, Marketing, and Strategic Management. A recipient of prestigious awards, he also serves on editorial boards and professional bodies, contributing significantly to academia, industry collaboration, and student-centered learning.

R. Vettriselvan, an Associate Professor at AMET University, specializes in HRM and Marketing. He is acting as a mentor at Saraswathi Institute of Medical Sciences, Hapur. He served as a Review Board member of the National Council for Higher Education, Malawi; Head of the Department of Commerce and Management Studies at St. Eugene University, Zambia; and Director of Research and Publication at St. John the Baptist University, Malawi. He has published 30 books and 100+ articles in peer-reviewed journals and edited book chapters. Under his guidance, two have completed and sevean are pursuing their PhD.

R. Ramya is a seasoned healthcare administrator dedicated to enhancing public health systems and empowering communities through outreach programs. With extensive experience and academic qualifications in Public Administration and Hospital Management, she serves as Vice Chairperson at Saraswathi Institute of Medical Sciences, Hapur driving excellence in healthcare delivery and education. Through her leadership and initiatives via charitable trusts, she addresses healthcare disparities and promotes community welfare, embodying a commitment to social impact and public health advancement.

Ir. Djoko Soelistya, M.M, CPHCM, CHRMP, Born in Surabaya, September 8, 1967, a practitioner who graduated with a Bachelor's degree in Civil Engineering from the Adhi Tama Surabaya Institute of Technology (ITATS), a Master's degree in Management from WR University. Supratman Surabaya and a Doctoral degree in Best Graduate from Airlangga University Surabaya in 2016, currently works as a Postgraduate Lecturer, Master's degree in Management study program at Muhammadiyah University of Gresik. The practitioner's background has been studied since 1989, and after a career in the Company he started teaching at the University since 2008 to 2024 (present), in the Management course group, especially the field of Human Resource Management, he also teaches at the Faculty of Economics and Business, Muhammadiyah University of Gresik (UMG), at Muhammadiyah University of Surabaya (UM Surabaya) and as a lecturer at the Open University (UT).

Index

www.ingramcontent.com/pod-product-compliance
Lightning Source LLC
LaVergne TN
LVHW081522300926
849161LV00006B/1263